D0108517

Equilibrium Thermodynamics

Second edition

Other titles in the European Physics Series

BRADDICK: Vibrations, Waves, and Diffraction
DUFFIN: Electricity and Magnetism
DUFFIN: Advanced Electricity and Magnetism
KIBBLE: Classical Mechanics
MATTHEWS: Introduction to Quantum Mechanics, Third Edition
MATTUCK: A Guide to Feynman Diagrams in the Many-Body Problem, Second Edition (in press)
SQUIRES: Practical Physics

Equilibrium Thermodynamics

Second edition

C. J. ADKINS

M.A., Ph.D., F.Inst.P.
Lecturer in Physics in the University of Cambridge and
Fellow of Jesus College, Cambridge

London · New York · St Louis · San Francisco · Sydney
Toronto · Mexico · Panama · Paris · Johannesburg · Singapore
Düsseldorf · Kuala Lumpur · New Delhi · São Paulo

Published by
McGRAW-HILL Book Company (UK) Limited
Maidenhead · Berkshire · England

Library of Congress Cataloging in Publication Data
Adkins, Clement John.
Equilibrium thermodynamics.
(European physics series)
Includes bibliographical references and index.
1. Thermodynamics. I. Title.
QC311.A3 1975 536′.7 74-34392
ISBN 0-07-084057-1

123456789–MAMM–76

10 9 8 7 6 5 4 3 2

To Tessara

Preface

There are many ways of writing a textbook on thermodynamics because the subject is relevant to so many branches of science. This book is written for students of physics and presents a comprehensive undergraduate course in classical thermodynamics. My intention has been to achieve a clear exposition: to give an account of the subject which is easy to learn from.

Several books have appeared recently in which classical and statistical thermodynamics are developed side by side. Although it is essential that the relationship between the two be established at some stage of an undergraduate education, there is much to be said for first teaching classical thermodynamics alone, for an ability to use it well depends largely on knowing what it can achieve *without* appealing to the microscopic nature of things. On the other hand, while it may be an interesting intellectual exercise to present thermodynamics without reference to microscopic structure, it is certainly educationally very foolish to do so. In this book, I make free use of microscopic ideas to illuminate the subject and to bring out its relevance to modern physics. I do not include any statistical mechanics nor any irreversible thermodynamics. Hence the title: *Equilibrium Thermodynamics*.

In writing this book, one of my most difficult problems has been how to present the *second law*. The advocates of the traditional approach based on the classical statements of Clausius or Kelvin argue that these are such simple generalizations of everyday experience that the experimental basis for the law is clearly displayed, and it is easy to accept it. However, the essential function of the second law is to introduce entropy and to define its properties. To reach the point where this is done, one must proceed from the Kelvin or Clausius statement by a long chain of argument involving heat engines, cyclic processes, and the rest. The advantage that may be gained in giving insight into the functioning of thermal machines is certainly outweighed by the deviousness by which one arrives at entropy.

An approach which chooses the opposite extreme has been introduced by L. Tisza and used by H. B. Callen in his book. There, the existence and properties of entropy are stated in a set of axioms. This has the advantage of introducing entropy directly, but is too far removed from the experimental basis for my liking.

The statement of *Carathéodory*, on the other hand, is based directly on the essential physical facts and leads quickly to entropy; but its popularity has been restricted because the traditional development from it depends on a rather difficult mathematical theorem about linear differential forms. Recently, however, Buchdahl has shown that this may be dispensed with by introducing the idea of *empirical entropy*, the relationship of which to irreversibility is somewhat like that of empirical temperature to equilibrium. In chapter 6, I present this approach, but I develop the traditional treatment first in earlier chapters. This is not entirely for lack of courage, for to achieve a proper understanding of entropy necessarily requires a considerable amount of thought and some may find the more concentrated arguments of chapter 6 too strong a diet for a first meeting with the second law. But I hope it will not simply be passed by; for to reassemble the structure of the second law from the Carathéodory statement having first followed through the traditional arguments is a great help to a proper understanding of its nature. I hope that, having read chapter 6, many students will eventually be able to discard the traditional structure in favour of Buchdahl's.

At the end of the book, I have collected together a number of problems which I hope will prove both instructive and stimulating. Many of these are based on old Cambridge University examination questions and some also appear in *Cavendish Problems in Classical Physics* edited by A. B. Pippard and published by Cambridge University Press. I am grateful to the Council of the Senate of the University and to the University Press for permission to use them. I should also like to thank Messrs. McGraw-Hill for permission to draw on copyright material in *Heat and Thermodynamics* by M. W. Zemansky for Tables 2.1 and 8.2 and for problems 3.3, 5.4, 8.15, and 12.3.

It is unwise to write something of the nature of a textbook without drawing on the experience of those who have tackled the task before, and it would be impossible to acknowledge all those who have contributed indirectly to its making. Of the books I have referred to often, I have found particularly valuable *Heat and Thermodynamics* by M. W. Zemansky, *Classical Thermodynamics* by A. B. Pippard, and *Heat, Thermodynamics and Statistical Physics* by F. H. Crawford; and Wilks' short book, *The Third Law of Thermody-*

namics, was a great help as I set about writing chapter 12. I also owe much to other members of this Laboratory and, in particular, must thank Dr J. Ashmead and Professor A. B. Pippard for helpful comments and profitable arguments. I am also most grateful to the members of the Department of Physics at Rutgers' University whose hospitality I enjoyed during the year that I was finishing the writing of this book. Finally, I must thank Susan Robbins, Jane Edwards, and Conni Slover for typing the manuscript and for their endless patience in trying to understand my writing.

C. J. Adkins
Cambridge, 1968

Preface to Second Edition

The most pressing reason for preparing this new edition was the need to bring the text of *Equilibrium Thermodynamics* up to date. In this respect, the most important changes concern units and those parts of chapter 2 dealing with scales of temperature. SI units are now used throughout, and the recommendations of the Symbols Committee of the Royal Society have generally been followed for the choice of symbols and the convention for showing units. The sections on temperature scales have been rewritten to bring the account into line with the current recommendations of the International Committee of Weights and Measures, and a new section, describing the International Practical Temperature Scale, has been added. While taking care to preserve the character of the first edition, I have also taken the opportunity of making various revisions throughout the text where it seemed that discussion could be improved.

I should like to thank undergraduate readers and colleagues in this and other universities for their questions, comments, and suggestions. If this new edition is judged an improvement on the old, some of the credit must go to them.

C. J. Adkins
Cambridge, 1975

Contents

1. Introduction

1.1. The Origins of Thermodynamics

The increase of mechanization during the nineteenth century involved the construction of machines, such as the steam engine, for the conversion of heat energy into mechanical power. It was from the study of these heat engines that thermodynamics grew. The initial development was rapid. By 1900, the subject was firmly established, and although its application had at first been restricted to thermal engineering, its laws were soon recognized to be of such great generality as to be useful and important in many other branches of science also. Broadly speaking, thermodynamics is applicable to all processes in which temperature or heat play an important part. In physics, it provides a way of understanding phenomena as different as thermal radiation on the one hand and the low temperature properties of paramagnetic salts on the other. It supplies the basic theory of chemical reactions and underlies much of chemical engineering. It is applicable not only to steam engines but to refrigeration and rocketry.

With this very wide range of application, it is possible to adopt various terms of reference within which to develop the subject. We shall choose examples which are primarily of interest to the physicist. The fundamental structure of the subject, however, is little affected by the applications one has in mind. This is because the basic theory can be developed in a precise and self-contained way with much of the rigour of a mathematical argument. To some this makes the subject seem too abstract and difficult, but we shall try to avoid this impression by developing the theory in the context of its applications to real physical systems.

1.2. The Macroscopic Approach

Thermodynamics sets out to describe and correlate the directly observable properties of substances: the volume of a gas, the expansion of a wire, the polarization of a dielectric. These are all *macroscopic* quantities, properties of the materials in bulk. If we were interested in the pressure exerted by a gas on the walls of its container, we could, in principle, adopt the *microscopic* approach and start from the equations of motion of the individual molecules, examine the statistics of their motions and finally arrive at an expression for the macroscopic quantity, pressure, in terms of momentum exchange at the boundaries of the container. But for many purposes an analysis in microscopic terms is unnecessary. The laws of thermodynamics enable us to interrelate the macroscopic quantities without making any microscopic assumptions at all. The great generality of thermodynamics is a direct consequence of this. By avoiding commitment to any particular microscopic interpretation, thermodynamics is not limited to particular applications nor subject to the fashions of microscopic theory.

On the other hand, it is possible to associate particular sorts of macroscopic behaviour with certain general kinds of microscopic change: an anomaly in a specific heat, for instance, may result from a change of atomic ordering in a crystal. However, since no microscopic assumptions are built into the thermodynamics it is never possible to *identify* a microscopic process by thermodynamic reasoning alone.

It is perhaps because thermodynamics is not concerned with fundamentals in the microscopic sense that it sometimes does not appeal readily to the physicist; but he will disregard it at his peril. It is precisely because it avoids microscopic theories that it is so valuable. It often yields answers to problems where an understanding of the fundamental processes involved might be difficult or impossible. It helps prevent mistakes; for any result which does not satisfy the requirements of thermodynamics must be wrong. But perhaps more important, a physicist's training is not only concerned with learning fundamental theories but also with developing a sensibility to the way in which physical systems behave, and here thermodynamics has a peculiar contribution to make by providing a very general framework of ideas from which the understanding of particular systems may more readily be achieved.

1.3. The Role of the Laws

In seeking to derive relationships between directly observable quantities, thermodynamics is essentially formulating rules which these quantities must obey under given conditions. These may apply to a substance undergoing a particular process, or they may be transformation rules which are useful in relating quantities which might be difficult to measure to ones which are more easily measured. For example, we will see that the ratio of the isothermal to the adiabatic compressibility is equal to the ratio of the principal specific heats:

$$\frac{\kappa_T}{\kappa_S}=\gamma=\frac{c_p}{c_V}.$$

Such a relation follows logically from the laws of thermodynamics. If they are true, then this relationship must always be true. If an experiment gave some other result, then there would be something wrong with the experiment, for else the whole structure of thermodynamics would collapse.

In order to derive these results in as simple a way as possible, it becomes necessary to define many new functions and concepts such as temperature, internal energy, specific heat, and entropy. With a given mass of gas, for example, we find that we usually only need specify its pressure and its volume to define its state precisely. These are *direct observables*. But if we wish to describe how the pressure and volume will change if that gas flows down a tube of varying cross-section, it is convenient to introduce a quantity called *enthalpy* which is conserved in the process. These new and more abstract quantities enable us to characterize processes or conditions in a simple way. They might be constants in a given process or they might take some extremal value under given conditions. Having defined these new quantities, we must, of course, expect them to be related to each other and to the direct observables in a way which follows logically from their definitions.

Of these new concepts, three are fundamental. Each follows from one of the laws of thermodynamics. From the zeroth law we are able to give a precise meaning to *temperature*. From the first law we are able to define *internal energy*, and from the second, *entropy*.

1.4. Systems, Surroundings, and Boundaries

A thermodynamic *system* is that portion of the universe which we select for investigation. A system may be simple or complex; it may be homogeneous or it may consist of many parts. A gas in a cylinder is a simple system. A mixture of phenol and water is a more complicated one: it contains two different substances or *components*, and for certain concentrations and temperatures separates into two phases. We define a *phase* as a system or part of a system which is homogeneous and has definite boundaries. A phase may be a chemically pure substance, or it may contain more than one component as is the case with the phenol–water mixture.

When it separates, both components are present in both phases, but in different concentrations.

Everything outside the system is called the *surroundings*, and the system is separated from the surroundings by its *boundary* (Fig. 1.1). In many cases, the boundary of a system may simply be its surface, as with a drop of liquid; but it is often convenient to contain the system within *walls* of some special kind that allow or prevent various sorts of interaction between the system and its surroundings. When we come to consider how a system may interact with its surroundings we shall find that interactions may be divided into

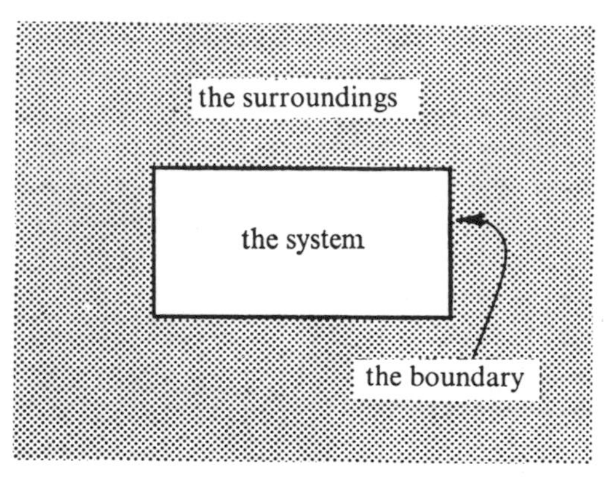

Fig. 1 . 1. A thermodynamic system.

two kinds. We may influence a system by doing *work* on it, or we may influence it *thermally*. Compression and magnetization are examples of work-like processes while heating in the flame of a bunsen burner is an example of a thermal process. Then a *rigid* wall prevents a system from changing its volume or shape so that no work of a mechanical nature may be done on it. Walls which prevent any thermal interaction are called *adiabatic*[1] walls, and a system enclosed in adiabatic walls is said to be *thermally isolated*. Such a system cannot exchange heat with its surroundings, but it may be possible to do work on it. Any changes which it undergoes will be *adiabatic*[1] *changes*. A Dewar vessel provides a good approximation to adiabatic walls. A wall which prevents interaction of *any* kind between the system and its surroundings is called an *isolating* wall, and the system is then said to be (*completely*) *isolated*. Walls which are not adiabatic (but through which a system may be influenced

[1] *Adiathermal* is also sometimes used.

thermally) are called *diathermal* and two systems separated by a diathermal wall are said to be in *thermal contact*.

In discussing chemical systems which contain different components, it is sometimes convenient to have a section of wall through which one or more of the components may pass while others are contained. Such a wall is said to be *semipermeable*. Hot quartz is permeable to helium but impermeable to other gases.

A system much used in developing the basic theory of thermodynamics is that of a gas contained in a smooth cylinder with a frictionless leakproof piston. This is a particularly helpful model to use as it is easy to visualize how changes take place and how the fundamental thermodynamic parameters of pressure and volume may be varied.

1.5. Thermodynamic Variables

The thermodynamic variables comprise the direct observables and the 'new' quantities discussed in section 1.3. They may be divided into two classes. Those of the first class are essentially local in character and include such quantities as pressure, electric field, force, and density.[2] They are known as the *intensive* variables.

[2] These quantities are not strictly local in the sense that it is possible to define them at a point. For example, we would define the local pressure in a gas by

$$p = \lim_{a \to 0} \frac{F}{a}$$

where F is the normal force exerted across a small area a. However, when $a \lesssim l^2$, where l is the mean free path of the gas molecules, the discrete nature of the molecular impacts becomes apparent and this quantity will fluctuate violently. As a becomes smaller, it becomes necessary to average over longer and longer times in order to achieve any similarity between the pressure as we have defined it and its macroscopic counterpart. A similar restriction applies to the other intensive variables. In the case of the electric field, the limit is set by the uncertainty principle, for we would define the local electric field by

$$E = \lim_{V \to 0} \frac{\dot{p}}{e}$$

where $\dot{p}$ is the rate of change of the momentum of a particle with charge e which we confine in the volume V. As we make V smaller so as to define E more nearly at a point, we eventually introduce a large uncertainty in the momentum of the particle by the restriction $\Delta p \, V^{1/3} \sim \hbar$, and it becomes impossible to observe $\dot{p}$. Fortunately, these restrictions do not concern us here, because by adopting the macroscopic approach in thermodynamics we can never hope to use it to describe systems on the atomic or quantum scale. Indeed, it is precisely because macroscopic quantities cease to have meaning that we cannot do so.

Those of the second class correspond to some measure of the system as a whole and include such quantities as mass, volume, internal energy, and length. These are proportional to the mass of the system if the intensive variables are kept constant, and for this reason are known as the *extensive* variables.

It is often convenient to refer to extensive quantities in terms of their values *per unit mass* of the system. These are known as *specific* variables. Generally, extensive variables are represented by capital letters and the derived specific quantities by the corresponding small letter. Thus, the volume of unit mass is called the specific volume and is given the symbol v.

Another useful convention is to add a suffix m to an extensive quantity when the amount of substance referred to is one mole. Thus, C_p is the heat capacity at constant pressure (unit, J K^{-1}) and C_{mp} is the molar heat capacity at constant pressure (unit, J K^{-1} mol^{-1}). The suffix is frequently dropped if there is no danger of confusion.

Many of the direct observables form conjugate pairs such that their product has the dimensions of energy. For these, the intensive member of each pair has the character of a force, and the extensive member that of a displacement. Some of these are listed in Table 1.1 together with the kind of system to which they particularly apply.

Any quantity which takes a unique value for each state of a system is called a *function of state*. The direct observables are obviously functions of state. In principle, it must be possible to

Table 1.1 Some Conjugate Pairs of Thermodynamic Variables

System	*Intensive variable*	*Extensive variable*
Fluid	pressure, p	volume, V
Filament	tensional force, $\mathscr{F}$	length, L
Film	surface tension, γ	area, A
Magnetic material	flux density, B	magnetic dipole moment, m
Dielectric	electric field, E	electric dipole moment[3], p
All systems	temperature, T	entropy, S
Generalized	force, X	displacement, x

[3] We prefer to use the conventional symbol p here despite the occasional possibility of confusion with pressure when problems involve both variables. In practice, the significance of the symbol is usually obvious from the dimensions of the quantities with which it appears.

express any function of state in terms of any set of variables which is sufficient to define the state of the system.

1.6. Thermodynamic Equilibrium

When a system suffers a change in its surroundings, it will usually be seen to undergo change. If the bulb of a thermometer is placed in a beaker of warm water, the mercury will begin to expand and will start to rise in the capillary. After a time, however, the system will be found to reach a state where no further change takes place and it is then said to have come to *thermodynamic equilibrium*. In general, the approach to thermodynamic equilibrium will involve both thermal and work-like interactions with the surroundings.

Similarly, if we place two systems in thermal contact, we generally find that changes will occur in both. When there is no longer any change (each has reached a state of thermodynamic equilibrium) the two systems are said to be in *thermal equilibrium*. In this case, we have prevented work-like interaction and allowed thermal interaction only. We shall eventually describe such a situation by saying that heat flows from one system to the other until they are at the same temperature.

Like mechanics, thermodynamics knows several kinds of equilibrium, and we define the stability of the equilibrium in a similar way. Thus a system is said to be in *stable* equilibrium if, after being slightly displaced, it returns to its original state. A system is in *metastable* equilibrium if it is stable for small displacements but unstable for larger ones. Certain systems also exhibit *neutral* equilibrium. Such a system may be displaced but will remain in the displaced condition when released. If a system is unstable to infinitesimal displacements it is said to be in *unstable* equilibrium.

It is worth pointing out that in the strictest sense neither mechanics nor thermodynamics knows truly unstable equilibrium, and the reason in both cases is the same. Equilibrium is defined in terms of macroscopic variables which are large scale averages of quantities which, on the microscopic scale, are subject to fluctuations. Pressure exerted by a gas is the macroscopic average of the impulses from discrete molecular impacts. The atoms of a solid are always in thermal motion. Although in large systems these fluctuations may be relatively unimportant, any fluctuation, however small, is, by definition, sufficient to destroy unstable equilibrium. Thus no truly unstable equilibrium exists although in some systems, the size of the displacement for which the system remains in metastable equilibrium may be so small that the system is loosely spoken of as being unstable. Being defined in terms of macroscopic quantities, equilibrium is itself a macroscopic concept. We may only apply the idea of equilibrium to large bodies, to systems of many particles. The Brownian movement of a colloid particle shows that it is certainly not in equilibrium. On the other hand, the mean density of colloid particles at different heights in a suspension does obey rules which may be derived from our ideas of equilibrium. Some examples of the different kinds of equilibrium are illustrated in Fig. 1.2.

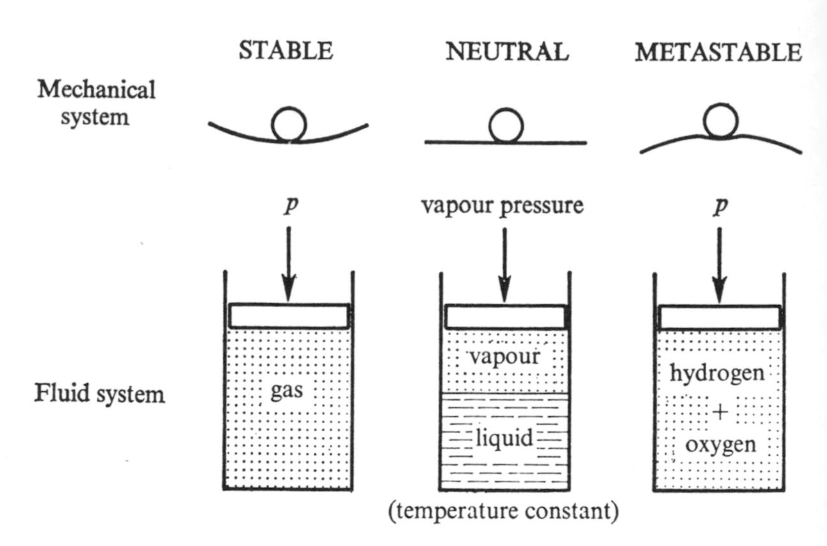

Fig. 1.2. Different kinds of equilibrium.

Stable equilibrium: A ball in a depression and a cylinder of gas at constant temperature will both eventually return to their initial states if displaced and released.

Neutral equilibrium: A ball on a horizontal plane may be displaced to any position on the plane and will remain there when released. Similarly, a system consisting of a liquid and its vapour at constant temperature also shows neutral equilibrium, for the vapour pressure depends on temperature only so that as long as both phases are present, change of volume simply causes condensation or vaporization without change of pressure, and the system remains in equilibrium with its surroundings.

Metastable equilibrium: A ball in a small hollow on an otherwise convex surface is only stable to small displacements. A mixture of hydrogen and oxygen in a thermally isolated vessel is also stable to small displacements but a large compression can raise the temperature sufficiently for the mixture to explode.

1.7. Thermodynamic Reversibility

When a system undergoes a series of changes a thermodynamic *process* is said to take place. A process is said to be reversible if, and only if, its direction can be reversed by an *infinitesimal* change in the conditions. It is not enough if it may only be reversed by a finite change. Thermodynamic reversibility requires two conditions to be satisfied: the process must be *quasistatic* and there must be no *hysteresis*.

Quasistatic Processes. To be quasistatic, a process must be carried out so slowly that every state through which the system passes may be considered an equilibrium state. Strictly speaking, this means that the process should be carried out infinitely slowly. Fast changes cause disequilibrium between different parts of a system. For example, suppose a gas is to be compressed from the state (p_1, V_1) to the state (p_2, V_2) (Fig. 1.3). If the compression is

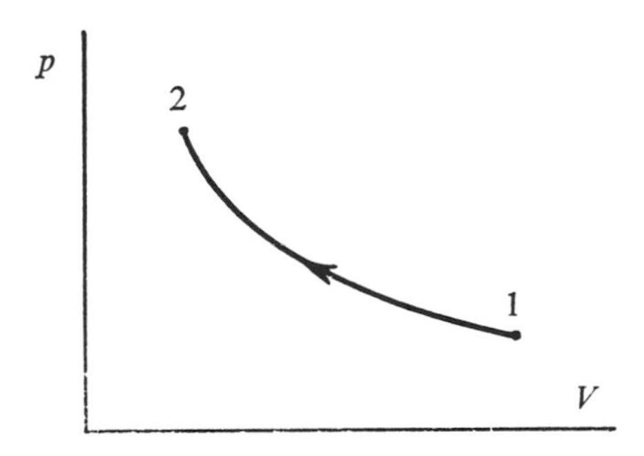

Fig. 1.3. The indicator diagram for a reversible process in a simple fluid.

performed sufficiently slowly, there will always be time for the gas to adjust to the changing environment and it will never depart significantly from equilibrium with it. Every state which the gas passes through will be an equilibrium state, and, clearly, the process may be reversed at any time by reversing the operations on the system. If, instead, we compress the gas rapidly by a sudden movement of the piston, sound waves or shock waves will be set up in the gas creating regions of different pressure and temperature. Clearly, such a change is not reversible. (We cannot extract the sound waves by moving the piston out again.)

Since an equilibrium state corresponds to definite values of the system parameters, we may represent a quasistatic process by plotting how the parameters vary as the system passes from the initial to the final state (Fig. 1.3). Such a representation is known as an *indicator diagram*. In a non-quasistatic change, system parameters do not define the states through which the system passes nor can they describe the processes it undergoes. Non-quasistatic processes should therefore not be represented by a line on an indicator diagram.

Hysteresis. When a process is reversed in a system with hysteresis, the system does not retrace its previous path, but proceeds by a different one. A common example is found in the magnetization of iron. (Fig. 1.4.) If carried out sufficiently slowly, each state through which the system passes may be considered as an equilibrium state. The variables are, at all times, well defined and the process may be represented on an indicator diagram. Nevertheless, here also, it is clear that the system parameters do not uniquely define the state of the system since their relationship depends on the previous history of the system. Friction is a common cause of hysteresis.[4]

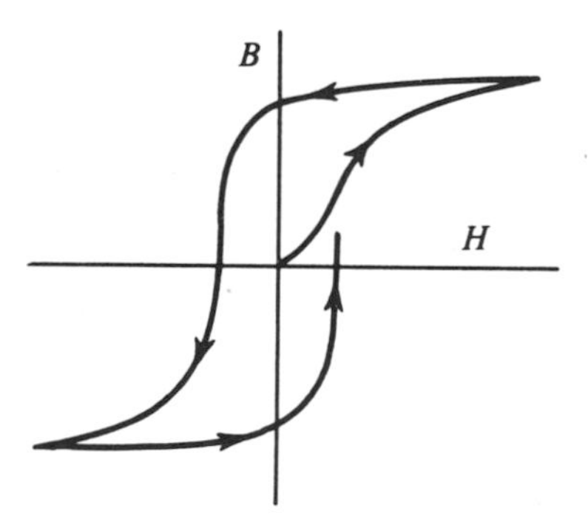

Fig. 1.4. An example of hysteresis: the magnetization of iron.

1.8. Degrees of Freedom

For any particular system we may list many thermodynamic variables which describe different aspects of its properties. If they are functions of state, the values of these variables will be determined by the state of the system: for a particular state, they will have a particular set of values. Many of them, however, will be related to one another in some way. For example, density is the ratio of mass to volume. We may therefore ask, what is the minimum number of variables whose values must be specified in order that the state of the system be uniquely determined? What is the number of independent variables? How many *degrees of freedom* does the system have?

The answer to this question obviously depends on the nature of the system and on how many conditions or constraints we impose on it. Very often, for a simple system, it is possible to say immediately how many degrees of freedom it has from a knowledge of its properties. A wire subject to tension only, for example, has two degrees of freedom, for we know that its length depends on the temperature and tension only. In other cases, it may help to make two lists, one of the system's variables and one of the conditions which they must satisfy. The number of degrees of freedom is then the number of variables, n_V, less the number of independent conditions, n_C. For a given mass of a simple gas, we might draw up the following lists:

Variables		*Conditions*	
pressure,	p	$M = \text{constant}$	
volume,	V	$\rho = M/V$	
temperature,	T	$p = p(V, T)$	(the gas law)
mass,	M		
density,	ρ		
$n_V = 5$		$n_C = 3$	

$$\mathcal{N} = n_V - n_C = 2.$$

There are two degrees of freedom, for which we may choose any convenient pair of the variables. It does not matter if we include

[4] Strictly speaking, the distinction between a non-quasistatic process and a hysteretic process is only one of time scale. If we place a piece of iron in a magnetic field there is a unique state of the iron corresponding to the lowest energy configuration of the system. As soon as the field is applied, we are really placing the iron in a *metastable* state from which it would have to proceed to the truly stable state by a series of minute changes on a molecular scale. The potential barriers between these steps, however, are so large that the approach to equilibrium only proceeds at an extremely slow rate, one which is quite negligible on any normal time scale. Since a hysteretic process is, in this sense, a non-quasistatic one on an enormous time scale, we might expect there to be inhomogeneities within the system as with normal non-quasistatic processes. From our knowledge of ferromagnetism we know this to be the case. The irregularities in the motion of domain

too many variables in the first list, because any new variable will introduce a corresponding constraint in the form of an equation relating it to others already listed. If we had included the compressibility of the gas, for example, we would also have had the equation defining it in terms of the pressure and volume:

$$\kappa_T = -\frac{1}{V}\left(\frac{\partial V}{\partial p}\right)_T.$$

The presence of more than one phase in a system or of several components obviously introduces more variables, but we will postpone the detailed discussion of such systems until we have found how to express the condition for them to be in thermodynamic equilibrium. (Chapters 10 and 11.)

In general, it is found that the number of degrees of freedom possessed by a system of given composition, *including its total mass*, is given by

▶ $$\mathcal{N} = 2 + n_W - n_C$$

where n_W is the number of ways in which work may be done on the system and n_C is the number of conditions to which it is subjected. (Mass = constant, temperature = constant, etc.)

Very often, certain coordinates play no significant part in the physical processes under examination, and it is then possible to disregard them. We know, for instance, that when a wire is stretched it suffers a very small change in volume, and that therefore work will be done on it by the hydrostatic pressure of any surrounding fluid. Strictly, the wire requires three parameters to define its state, say tension $\mathcal{F}$, length L, and volume V; but if the hydrostatic pressure is small, the work done by it may be negligible in comparison with the work done by the tensional force so that volume changes play no significant part in determining the simple elastic behaviour. It is then only necessary to retain $\mathcal{F}$ and L.

We shall later use various functions known as the thermodynamic potentials. (Chapters 7 *et seq.*) It is worth mentioning in advance that the expressions for the differentials of these functions contain

walls, which may be observed under a microscope, indicate the presence of inhomogeneities which hinder magnetic rearrangement. It is precisely these inhomogeneities which prevent the attainment of the true equilibrium state within any normal period of time. Since, by and large, the times taken by thermodynamic systems to reach true equilibrium tend to one or other of the extremes, it is convenient to retain the distinction between the non-quasistatic and hysteretic processes.

several terms, and that the number of these terms is equal to the number of degrees of freedom of the system. (Clearly, this must be the case in the expression for the differential of any function of state.) However, for the moment, these details need not concern us, for we shall develop the basic ideas of thermodynamics by referring only to very simple systems.

1.9. Some Useful Mathematical Results

Differential coefficients relating the rate of change of one thermodynamic variable with another are very important in thermodynamics. They are known as *thermodynamic coefficients*, and since their manipulation is a vital part of thermodynamic calculation it is important to understand their meaning and to be familiar with some basic mathematical results which are of help in handling them.

1.9.1. The Reciprocal and Reciprocity Theorems

Suppose that three variables are related:

$$F\,(x, y, z)=0.$$

Then, in principle, this equation may be rearranged to express one of the variables in terms of the other two as independent variables:

$$x=x\,(y, z).$$

Differentiating by parts, we have

$$\mathrm{d}x=\left(\frac{\partial x}{\partial y}\right)_z \mathrm{d}y+\left(\frac{\partial x}{\partial z}\right)_y \mathrm{d}z \tag{1.1}$$

where the terms in brackets are the *partial differentials* of x. Formally, the partial differentials are defined analogously to normal differentials:

$$\left(\frac{\partial x}{\partial y}\right)_z=\underset{\delta y\to 0}{\mathrm{Lt}}\;\frac{x(y+\delta y, z)-x(y, z)}{\delta y}. \tag{1.2}$$

Equation (1.1) expresses the change in x which results from changes in both of the independent variables on which x depends. We may write an analogous equation for $\mathrm{d}z$:

$$\mathrm{d}z=\left(\frac{\partial z}{\partial x}\right)_y \mathrm{d}x+\left(\frac{\partial z}{\partial y}\right)_x \mathrm{d}y.$$

Substituting this in (1.1) we obtain

$$\mathrm{d}x = \left(\frac{\partial x}{\partial z}\right)_y \left(\frac{\partial z}{\partial x}\right)_y \mathrm{d}x + \left[\left(\frac{\partial x}{\partial y}\right)_z + \left(\frac{\partial x}{\partial z}\right)_y \left(\frac{\partial z}{\partial y}\right)_x\right] \mathrm{d}y. \qquad (1.3)$$

This result must of course always be true whichever pair of variables we choose to think of as independent. In particular, we could choose x and y as independent. Then, if we make $\mathrm{d}y=0$ and $\mathrm{d}x\neq 0$, (1.3) gives

$$\left(\frac{\partial x}{\partial z}\right)_y \left(\frac{\partial z}{\partial x}\right)_y = 1,$$

or,

▶ $$\left(\frac{\partial x}{\partial z}\right)_y = \frac{1}{\left(\frac{\partial z}{\partial x}\right)_y}. \qquad (1.4)$$

This is the *reciprocal theorem* which allows us to replace any partial derivative by the reciprocal of the inverted derivative *with the same variable held constant.*

If we now substitute in (1.3) $\mathrm{d}x=0$ and $\mathrm{d}y\neq 0$ the term in brackets must be identically zero giving

$$\left(\frac{\partial x}{\partial y}\right)_z = -\left(\frac{\partial x}{\partial z}\right)_y \left(\frac{\partial z}{\partial y}\right)_x,$$

or,

▶ $$\left(\frac{\partial x}{\partial y}\right)_z \left(\frac{\partial y}{\partial z}\right)_x \left(\frac{\partial z}{\partial x}\right)_y = -1.$$

This is the *reciprocity theorem*. It may be written starting with any derivative then following through the other variables in cyclic order. (This gives a dimensionless combination.) This relation is most often used to split up a derivative into a product of more convenient derivatives.

1.9.2. Order of Differentiation

Second and higher order derivatives may involve differentiation with respect to more than one of the independent variables. It is a general result that the derivative does not depend on the order of differentiation. We may easily show this for a second order differential.

Suppose that $x=x(y, z)$. For given small changes in y and z we may find the change in x by expanding x in a Taylor's series in each variable in turn. We obtain the result we require by comparing the expressions we get by going from the initial to the final values of y and z by two different paths. These are illustrated in Fig. 1.5.

Proceeding first from the initial values of the independent variables represented by the point 1 to A by changing y, we have, by Taylor's theorem,

$$x_A = x_1 + \left(\frac{\partial x_1}{\partial y}\right)_z \delta y + \tfrac{1}{2}\left(\frac{\partial^2 x_1}{\partial y^2}\right)_z (\delta y)^2 + \ldots, \tag{1.5}$$

where the suffixes inside the brackets indicate the point at which the differentials are evaluated.

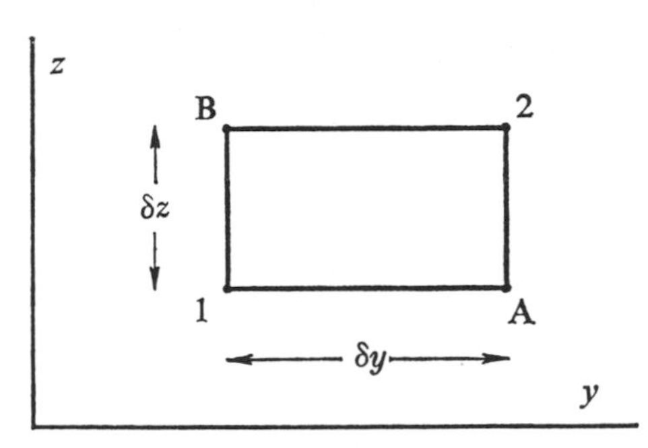

Fig. 1.5. Calculating the change in the function $x(y, z)$.

Now, proceeding from A to 2,

$$x_2 = x_A + \left(\frac{\partial x_A}{\partial z}\right)_y \delta z + \tfrac{1}{2}\left(\frac{\partial^2 x_A}{\partial z^2}\right)_y (\delta z)^2 + \ldots. \tag{1.6}$$

Substituting (1.5) in (1.6), and retaining terms up to second order only,

$$x_2 = x_1 + \left(\frac{\partial x_1}{\partial y}\right)_z \delta y + \left(\frac{\partial x_1}{\partial z}\right)_y \delta z + \tfrac{1}{2}\left(\frac{\partial^2 x_1}{\partial y^2}\right)_z (\delta y)^2$$
$$+ \tfrac{1}{2}\left(\frac{\partial^2 x_1}{\partial z^2}\right)_y (\delta z)^2 + \frac{\partial}{\partial z}\left(\frac{\partial x_1}{\partial y}\right)_z \delta y \delta z + \ldots. \tag{1.7}$$

If we proceed via B, the only difference is in the last term:

$$x_2 = x_1 + \left(\frac{\partial x_1}{\partial y}\right)_z \delta y + \left(\frac{\partial x_1}{\partial z}\right)_y \delta z + \tfrac{1}{2}\left(\frac{\partial^2 x_1}{\partial y^2}\right)_z (\delta y)^2$$
$$+ \tfrac{1}{2}\left(\frac{\partial^2 x_1}{\partial z^2}\right)_y (\delta z)^2 + \frac{\partial}{\partial y}\left(\frac{\partial x_1}{\partial z}\right)_y \delta z \delta y + \ldots. \tag{1.8}$$

Clearly, whichever way x_2 is calculated, the result must be the same so that the last two terms must be equal:

$$\frac{\partial}{\partial z}\left(\frac{\partial x}{\partial y}\right)_z=\frac{\partial}{\partial y}\left(\frac{\partial x}{\partial z}\right)_y,$$

or,

$$\blacktriangleright \qquad \frac{\partial^2 x}{\partial z\,\partial y}=\frac{\partial^2 x}{\partial y\,\partial z}. \tag{1.9}$$

It should be noticed that if x_1 and x_2 are so close that terms above the first order may be neglected, equations (1.7) and (1.8) are identical to the simple differential form (1.1).

1.9.3. Exact Differentials

We have seen that if x is a function of y and z, it is always possible to write the infinitesimal change in x which results from infinitesimal changes in y and z in the differential form

$$\mathrm{d}x=Y\,\mathrm{d}y+Z\,\mathrm{d}z \tag{1.10}$$

where

$$Y=\left(\frac{\partial x}{\partial y}\right)_z$$

and

$$Z=\left(\frac{\partial x}{\partial z}\right)_y.$$

Since $\mathrm{d}x$ is the differential of a function of y and z, it may, in principle, always be integrated. For this reason it is known as an *exact* or *perfect* differential. Clearly, the differential of a function of state must always be exact since a function of state is, by definition, a single-valued function of the state variables.

If, in thermodynamics, an infinitesimal quantity is not the differential of a function of state, it is written $đx$, where the stroke on the d indicates that it is *inexact* and cannot in general be integrated. Sometimes, however, knowledge about *how* a change takes place, information about the path by which the system proceeds from its initial to its final state, makes it possible to integrate an inexact differential, but this always requires more information than is provided by a knowledge of the initial and final states alone.

Applying (1.9) to (1.10), we obtain

$$\blacktriangleright \qquad \left(\frac{\partial Y}{\partial z}\right)_y=\left(\frac{\partial Z}{\partial y}\right)_z. \tag{1.11}$$

It may be shown that (1.11) is a necessary and sufficient condition for $\mathrm{d}x$ to be exact.

2. The Zeroth Law

2.1. The Zeroth Law

The zeroth law of thermodynamics is concerned with the properties of systems in thermal equilibrium, and the concept of temperature follows directly from it. The statement of the law is as follows:

If two systems are separately in thermal equilibrium with a third, then they must also be in thermal equilibrium with each other.

The sort of experiment on which this law is based is illustrated in Fig. 2.1 where for our three systems we choose a mercury thermometer and two cylinders of gas. The zeroth law simply says that if there is no change when the thermometer is placed in thermal contact with system A nor when it is placed in thermal contact with system B, then there will be no change if systems A and B are placed in thermal contact with one another.

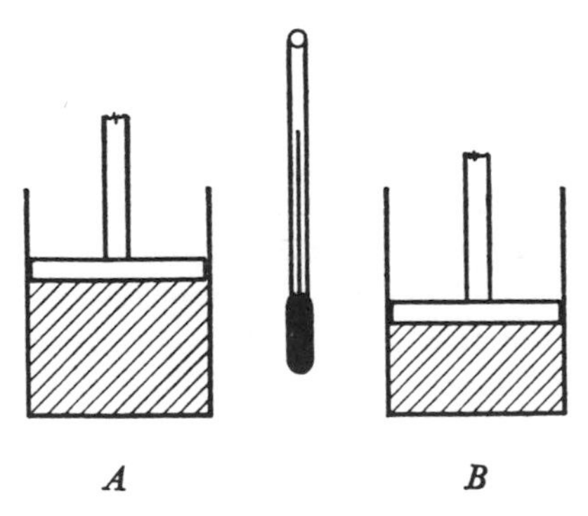

Fig. 2.1. An illustration of the zeroth law.

If one chooses to define temperature as the reading given by the mercury thermometer, then the zeroth law is only saying that if two bodies are at the same temperature, they will be in thermal equilibrium when placed in thermal contact. The connection between thermal equilibrium and temperature is trivial. However, it is possible to use the zeroth law to demonstrate the existence of temperature in a more general way. The argument does not follow from the zeroth law alone, for we shall also use our knowledge of the behaviour of real physical systems.

2.2. Temperature

We shall demonstrate the existence of temperature by applying the zeroth law to three systems, 1, 2, and 3, each of which consists of a certain mass of fluid enclosed in a cylinder fitted with the usual frictionless piston. For each system we choose as the parameters of state the pressure p, and the volume V. We use system 3 as a reference, setting it in a chosen state by adjusting the values of p_3 and V_3. Now, in principle, we know that by suitable manipulations we can obtain any values for p_1 and V_1 we choose. In the absence of constraints they are independent variables. However, we know

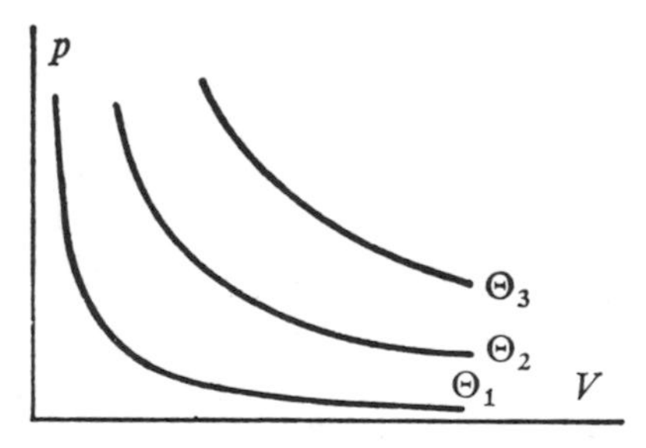

Fig. 2.2. Isotherms for a simple fluid.

from experiment, that if we demand that 1 be in thermal equilibrium with 3 then the new constraint leaves only one independent variable. That is, if we choose a particular value for p_1, then V_1 will be uniquely determined. Thus, by choosing a series of values for p_1 and determining the corresponding V_1's we may plot all the values of p_1 and V_1 which give thermal equilibrium with 3. Such a plot is an *isotherm*. To the isotherm and the corresponding reference state of 3 we may attach an identifying symbol, Θ_1. By choosing other reference states, we may plot out as many isotherms of system 1 as we choose, and to each we may attach a label: Θ_2, Θ_3, . . . , etc. (Fig. 2.2.) Using the same reference states of 3, corresponding isotherms can be constructed for 2. The zeroth law then states that systems 1 and 2 will be in thermal equilibrium at *any points on corresponding isotherms*. They must have some property in common which allows this to be so. This property is called temperature.

This argument may be presented more formally as follows. We create a reference state by fixing p_3 and V_3. Then, if we choose a particular value for p_1 and demand thermal equilibrium between systems 1 and 3, V_1 will be determined. That is, there must be a fixed relationship between the four variables. This may be expressed in the form

$$F_1(p_1, V_1, p_3, V_3)=0. \tag{2.1}$$

Equation (2.1) expresses the condition for equilibrium between 1 and 3. Similarly, for 2 and 3 to be in equilibrium we must have

$$F_2(p_2, V_2, p_3, V_3)=0. \tag{2.2}$$

These equations may, in principle, be solved for, say, p_3:

$$p_3=f_1(p_1, V_1, V_3) \tag{2.3}$$

$$p_3=f_2(p_2, V_2, V_3). \tag{2.4}$$

Eliminating p_3 from equations (2.3) and (2.4) we get

$$f_1(p_1, V_1, V_3)=f_2(p_2, V_2, V_3) \tag{2.5}$$

which may be solved for, say, p_1:

$$p_1=g(V_1, p_2, V_2, V_3). \tag{2.6}$$

But, by the zeroth law, if 1 and 2 are separately in equilibrium with 3 they must be in equilibrium with each other. This requires

$$F_3(p_1, V_1, p_2, V_2)=0.$$

Again, we may solve this for p_1:

$$p_1=f_3(V_1, p_2, V_2). \tag{2.7}$$

Now equation (2.7) states that p_1 is determined by the three variables V_1, p_2, and V_2 alone so that V_3 must cancel out in (2.6). Similarly, it must drop out of the equation in its earlier form, (2.5), also, so that (2.5) must really be of the form

$$\Phi_1(p_1, V_1)=\Phi_2(p_2, V_2). \tag{2.8}$$

Equation (2.8) expresses the condition for thermal equilibrium between systems 1 and 2. It shows that when two (or more) systems are in thermal equilibrium there is, for each one, a function of its parameters which takes a common value for all the systems. Thus, for any system in thermal equilibrium with a given reference system (i.e., at a given temperature) we may write

► $$\Phi(p, V)=\Theta \tag{2.9}$$

where Θ is the same for all systems. Equation (2.9) is called the *equation of state* and Θ is the *empirical temperature*.

2.3. Scales of Temperature

Temperature, as we have defined it, need not bear any simple relation to our intuitive ideas of hotness. Strictly speaking, we have done no more than construct isotherms and attach symbols to them. Clearly, it is desirable to define temperature in such a way that the temperatures form an ordered sequence corresponding to our ideas of hotness. This is what we do when we construct a *scale* of temperature.

To establish a particular empirical scale we select some system with suitable thermometric properties and adopt a convenient method of assigning numerical values for the temperatures of its isotherms. If the thermometric property we use is x, then the simplest possible procedure is to take the scale as linearly proportional to x:

$$\Theta(x) = ax.$$

We then fix the constant a either by choosing the *value* of the temperature at *one* reference point, or by choosing the *size* of the unit so that a given number of units lie between *two* fixed points. Either procedure will define a unique scale *for any one thermometer*, but measurements of a temperature made with different thermometers will not in general agree with one another because the temperature dependencies of the chosen thermometric properties may be quite different.

2.4. The Perfect Gas Scale

In the search for a scale which did not depend on the properties of a particular substance, it was found that disagreement was small among measurements based on the behaviour of gases. If we describe the state of a gas by the two parameters p and V, the simplest way of constructing a scale is to keep one of the variables fixed and to take the temperature as proportional to the other (Fig. 2.3). Historically, the constant was chosen to give 100 units between the ice point (the temperature at which ice melts at a pressure of one atmosphere) and the steam point (the temperature at which water boils at a pressure of one atmosphere). Thus, for a *constant pressure gas thermometer* we would have

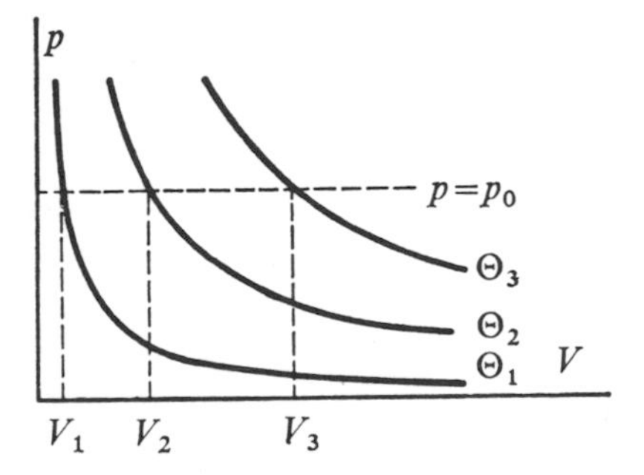

Fig. 2.3. Construction of a temperature scale based on the behaviour of a gas.

The simplest procedure is to keep one parameter constant and to take the temperature as linearly proportional to the other: $\Theta = aV + b$.

$$\Theta = \frac{100}{V_s - V_i} V \qquad (p = \text{const.}) \tag{2.10}$$

and for a *constant volume gas thermometer*

$$\Theta = \frac{100}{p_s - p_i} P \qquad (V = \text{const.}) \tag{2.11}$$

where the suffixes 's' and 'i' refer to the steam and ice points respectively. While temperature measurements made in this way are usually in reasonable agreement with one another, it was found that in the low density limit ($p \rightarrow 0$) *all* gases give the *same* value for a temperature. The temperature scale so defined was known as the *perfect gas absolute scale*. (The word 'absolute' here indicates that there is no shift of the zero by subtraction of a constant to bring the zero into the range of everyday temperatures, as is done for the Celsius scale which is discussed in section 2.7.)

2.5. Thermodynamic Temperature[1]

All gases give the same temperature scale in the low pressure limit, because in that limit their behaviour tends to that of the ideal or perfect gas. We shall discuss the reasons for this in section 8.3. In section 8.2 we shall show that the equation of state of the perfect gas is

$$pV = RT$$

[1] Previously known as the Kelvin scale.

where R is a constant and T is *thermodynamic temperature*, the fundamental measure of temperature which arises naturally in thermodynamic theory (section 4.6). From the form of this equation we see immediately why gases in the low pressure limit define a unique scale of temperature. In fact, gases in the ideal limit are quite exceptional in that thermodynamic temperature appears in such a simple way in their equation of state. This is why determination of thermodynamic temperature is almost always ultimately based on gas thermometry:

$$▶ \quad T = \lim_{p \to 0} (pV)/R \tag{2.12}$$

where the size of the unit is determined by the value of R. Originally R was chosen to give 100 units between the ice and steam points, but this requires measurements at *two* reference points. The experimental uncertainties are reduced if only one reference point is involved. Consequently, in 1954, the Tenth General Conference of Weights and Measures decided to adopt the other method of fixing the constant of proportionality, namely to specify the *value* of the thermodynamic temperature at one reference point.[2] Because of its greater reproducibility than, say, the ice point, the reference point chosen was the triple point of water—the temperature at which ice, water, and water vapour coexist in equilibrium. The value of thermodynamic temperature allotted to this was 273·16; the unit so defined is called the kelvin and is given the symbol K (no degree sign). Put differently: *the kelvin is the fraction* 1/273·16 *of the thermodynamic temperature of the triple point of water*. Thermodynamic temperatures determined by gas thermometry are therefore given by

$$▶ \quad T/\mathrm{K} = 273{\cdot}16 \, \frac{\lim_{p \to 0} (pV)}{\lim_{p \to 0} (pV)_{\mathrm{tr}}}. \tag{2.13}$$

The reason for the choice of 273·16 K for the triple point of water is that, to the accuracy of currently available measurements, this gives exactly 100 K for the difference of the thermodynamic temperatures of the ice and steam points. The ice point is 273·15 K and the steam point 373·15 K. With thermodynamic temperature so defined,

[2] It was Kelvin himself who first suggested that the size of the unit should be fixed in this way; hence it is appropriate that it should be named after him. See W. Thomson (Lord Kelvin), *Phil. Trans. Roy. Soc.* A, **144**, 350, 1854.

the temperature of the triple point of water is fixed by definition, but we may find, if measurement techniques improve, that the ice point is not exactly 273·15 K, that the steam point is not exactly 373·15 K, and that there are not exactly 100 K between them.

2.6. The Celsius Scale

For practical purposes it is convenient to have a scale whose zero is towards the bottom of the range of temperatures commonly encountered. That used is the *Celsius scale*, which is formally defined by

$$t/°\mathrm{C} = T/\mathrm{K} - 273{\cdot}15. \qquad (2.14)$$

The symbol t is always used for Celsius temperatures and T is reserved for thermodynamic temperature. The unit is the degree Celsius (°C, degree sign included), which is identical in size to the kelvin. On the Celsius scale, the triple point is 0·01 °C by definition and the ice and steam points 0 and 100 °C respectively.

2.7. Some Common Thermometers

For the reasons mentioned above, determinations of thermodynamic temperatures are usually ultimately based on gas thermometry; but gas thermometers are inconvenient and difficult to use, especially when much accuracy is required. Except when absolute determinations have to be made, other kinds of thermometers are generally used, the choice depending on such criteria as convenience or sensitivity rather than high absolute accuracy. However, if the *thermodynamic* temperature is required (and not just some convenient empirical scale), the practical thermometers have to be calibrated. To help with this, the thermodynamic temperatures of a number of primary and secondary reference points have been measured with great accuracy. Table 2.1 lists the standard and primary reference points and the temperatures assigned to them by the International Committee of Weights and Measures in 1968. The assigned values are given to greater precision than the estimated uncertainties because they are used as defining points for the International Practical Temperature Scale (see section 2.8).

Thermometers based on the *expansion of liquids* can cover a remarkable range of temperature. They are reasonably linear, but not very sensitive. If much accuracy is required, many corrections have to be applied. Some liquids which are commonly used are:

mercury	between	−39	and	+350°C
ethyl alcohol	between	−117	and	+78°C
pentane	between	−130	and	+36°C.

Suitable mixtures of the lighter paraffins extend the range to below −200°C.

Table 2.1 Thermodynamic and Celsius Temperatures of Standard and Primary Reference Points*

	T/K	t/°C	*Estimated uncertainty*/K
Standard			
Triple point of water†	273·16	0·01	exact by definition
Primary			
Triple point of equilibrium hydrogen‡	13·81	−259·34	0·01
Boiling point of equilibrium hydrogen at 25/76 atm pressure	17·042	−256·108	0·01
Boiling point of equilibrium hydrogen at 1 atm pressure	20·28	−252·87	0·01
Boiling point of neon at 1 atm pressure	27·102	−246·048	0·01
Triple point of oxygen	54·361	−218·789	0·01
Boiling point of oxygen at 1 atm pressure	90·188	−182·962	0·01
Boiling point of water at 1 atm pressure	373·15	100·00	0·005
Melting point of zinc at 1 atm pressure	692·73	419·58	0·03
Melting point of silver at 1 atm pressure	1235·08	961·93	0·2
Melting point of gold at 1 atm pressure	1337·58	1064·43	0·2

* For further information and details of secondary reference points, see *The International Practical Temperature Scale of 1968*, Her Majesty's Stationery Office, London, 1969.

† The water should have the isotopic composition of ocean water.

‡ Hydrogen has two molecular modifications, *ortho* and *para*. The equilibrium ratio of *ortho*- to *para*-hydrogen is temperature dependent, but conversion of one form to the other is normally slow unless a catalyst is present. In the present context *equilibrium hydrogen* means hydrogen in which the *ortho*/*para* ratio has its equilibrium value *for the temperature concerned*.

Resistance thermometers, based on the variation of electrical resistance of a metal with temperature, cover an even greater range.

Platinum is often used as it is comparatively easy to purify, purity improving its performance at low temperatures, and it also has a high melting point (1770°C). Between 70 K and 1200°C it is capable of very high accuracy. It is not far from linear, and for moderate accuracy a quadratic relation between the resistance and thermodynamic temperature, $R = R_0(1 + aT + bT^2)$, gives a good fit over the whole temperature range. For very high precision a cubic term is added below 0°C. Between 1200°C and the melting point, the thermometer is still useful although its accuracy falls off at the higher temperatures.

Thermocouples, thermometers using the variation of the Seebeck e.m.f. with temperature (see section 9.4), cover much the same temperature range. Several commonly used thermocouples are listed in Table 2.2. The e.m.f. is generally well represented by an expression of the form $\mathscr{E} = a_1(\delta t) + a_2(\delta t)^2 + a_3(\delta t)^3$, where δt is the temperature difference between the junctions. If one junction is kept at a constant temperature, then the temperature of the other is given by an expression of the form $\mathscr{E} = b_0 + b_1 t + b_2 t^2 + b_3 t^3$, requiring four fixed points. Over restricted ranges, the higher terms may be discarded. The small voltages which have to be measured make it difficult to use thermocouples for highly accurate work, and their sensitivities are very dependent on any variation in the purity or composition of the metals used. However, they have several merits. They can be made very small and will respond quickly to changes in temperature. They are very useful for measuring small differences in temperature, and when high accuracy is not required, they are also very simple to use.

Table 2.2 Some Common Thermocouples

Pair	*Approximate sensitivity*/μV K^{-1}	*Normal working range*/°C
copper–constantan(1)	40	−200 to + 300
iron–constantan	50	−200 to + 750
chromel(2)–alumel(3)	40	−200 to +1200
platinum–platinum/rhodium(4)	6	−200 to +1450

(1) Usually 60% Cu, 40% Ni
(2) 90% Ni, 10% Cr
(3) 95% Ni plus Al, Si, Mn.
(4) 90% Pt, 10% Rh.

Another kind of thermometer which may be used over a very large range of temperature is based on the temperature variation of the electrical conductivity of a crystalline *semiconductor*.[3] For electrical conduction to occur in a semiconductor, current carriers (electrons or holes) have to be excited out of states in which they cannot contribute to current flow, into states in which they are able to move through the crystal and carry charge. This excitation normally occurs thermally and results in a strong dependence of the conductivity on temperature. Over a limited range of temperature the conductivity is approximately exponentially dependent on temperature varying as $\exp(-\varepsilon/kT)$, where ε is a constant. If the semiconductor is chosen so that $\varepsilon > kT$, its resistance will depend strongly on temperature. ε must not be too big, or the conductivity becomes too small and difficult to measure; but materials may be made with values of ε such as to provide good semiconducting thermometers over the temperature range from well below 1 K to above 300°C. The variation of the conductivity is so strong that any one material can only be used easily over a limited range (say a factor of ten in $1/T$), but this extreme sensitivity is their great merit. It is not difficult to detect changes in temperature of 1 part in 10^5 giving sensitivities of about 10 μK at 1 K and 1 mK at room temperature. At low temperatures a suitable semiconductor provides one of the most sensitive and reproducible thermometers available.

Carbon resistors of the type used in electronics also form sensitive low temperature thermometers. They are more widely used than semiconducting thermometers because they are cheap and readily available. Their mechanism of electrical conduction is not fully understood, but, as with crystalline semiconductors, excitation processes are involved (for transfer of charge between the grains of graphite, for example) and it is probably these which give rise to the strong temperature dependence. Carbon resistance thermometers are useful below 20 K. Below about 10 K their sensitivities are similar to semiconducting thermometers, making it possible to measure to about 10 μK at a few kelvins. In both cases the ultimate sensitivity is set by Johnson noise[4] in the thermometer. This cannot be overcome by increasing the measuring current through the resistor, because this eventually results in too much power

[3] See, S. A. Friedberg, 'Semiconductors as Thermometers', in *Temperature, Its Measurement and Control in Science and Industry*, **2**, Reinhold, 1955.

[4] B. I. Bleaney and B. Bleaney, *Electricity and Magnetism*, chapter 16, Oxford University Press, 1965.

being dissipated in the thermometer which prevents it from following the temperature of its surroundings.

In the range from 5 K to somewhat below 1 K, liquid ^{4}He is used as a refrigerant. The lighter isotope, ^{3}He, has a lower boiling point and is useful between about 1 K and 0·3 K. In both cases the temperature can be found by measuring the *vapour pressure*. Very accurate vapour pressure tables, based ultimately on helium gas thermometer measurements, are available for both isotopes.[5] Secondary thermometers for use below 5 K are almost always calibrated by helium vapour pressure thermometry.

Carbon and semiconductor resistance thermometers may be used below 1 K, but extrapolation downwards from the helium range soon becomes very inaccurate and it becomes necessary to use a thermometer whose law is known, or can be found. The magnetic susceptibility of some paramagnetic salts varies with temperature below a few kelvins. The susceptibility may easily be measured, (usually it is done by determining the inductance of a soil surrounding a sample of the salt) and this provides the basis of *susceptibility thermometry*. As long as the temperature is not too low, the susceptibility χ is well described by Curie's law, $\chi = a/T$ where a is a constant, and temperatures may be found by extrapolation from the helium range. Although Curie's law eventually breaks down, these thermometers may be used for lower temperatures as long as their susceptibility remains appreciably temperature dependent. However, the determination of thermodynamic temperature then requires a calibration procedure which depends on the second law, and will be discussed in section 8.8.2. Cerium magnesium nitrate is particularly useful at low temperatures as it obeys Curie's law to a few per cent down to below 4 mK, and may be extrapolated to this temperature from the helium range.

None of the thermometers described above is useful far above the gold point (1064°C) and in this range *radiation pyrometers* are used. These are based on the measurement of the radiation emitted by a body when hot. Both the colour and the total amount of radiation change with temperature (see section 8.9) and both these properties are used in pyrometry. *Optical pyrometers* use the colour of the radiation emitted by the hot body and usually consist of some device for

[5] For ^{4}He: *Journal of Research of the National Bureau of Standards* (*Washington*), 1960, **64A**, 1.
For ^{3}He: *Journal of Research of the National Bureau of Standards* (*Washington*), 1964, **68A**, 547, 559, 567, and 579.

matching the colour of an electric lamp filament to that of the radiation. *Broad band radiation pyrometers* measure the power radiated in a range of wavelengths selected by filters and are based on Planck's law. (Section 8.9.5.) *Total radiation pyrometers* measure the total power emitted and are based on Stefan's law. (Section 8.9.4.)

The ranges over which the various kinds of thermometer operate are shown in Fig. 2.4. The choice for a particular application does not simply depend on the range of temperature to be measured, but also on the conditions under which the measurement has to be made. Such factors as size, speed of response, and sensitivity will often determine the ultimate choice. For a detailed discussion of the experimental techniques involved in thermometry the student should consult other texts.[6]

2.8. The International Practical Temperature Scale

The experimental difficulties of accurate measurement of thermodynamic temperature with gas thermometers make it necessary for laboratories and standards institutions to have available a set of convenient practical thermometers whose behaviour is known in sufficient detail for them to be used for accurate interpolation between basic reference points whose thermodynamic temperatures have been determined with precision. This is the *raison d'être* of the International Practical Temperature Scale. The International Committee of Weights and Measures

(*a*) selects a set of reference points and *assigns* to these points values of thermodynamic temperature in the light of best available measurements,
(*b*) selects a set of thermometers for interpolation between the reference points, and
(*c*) agrees on the interpolation procedures to be used.

The reference points with their assigned temperatures, together with the specified thermometers and interpolation procedures, establish

[6] For a general discussion of thermometry, see: J. K. Roberts and A. R. Miller, *Heat and Thermodynamics*, Blackie, 1960.

For a discussion of low temperature thermometry, see: A. C. Rose-Innes, *Low Temperature Laboratory Techniques*, English Universities Press, 1973.

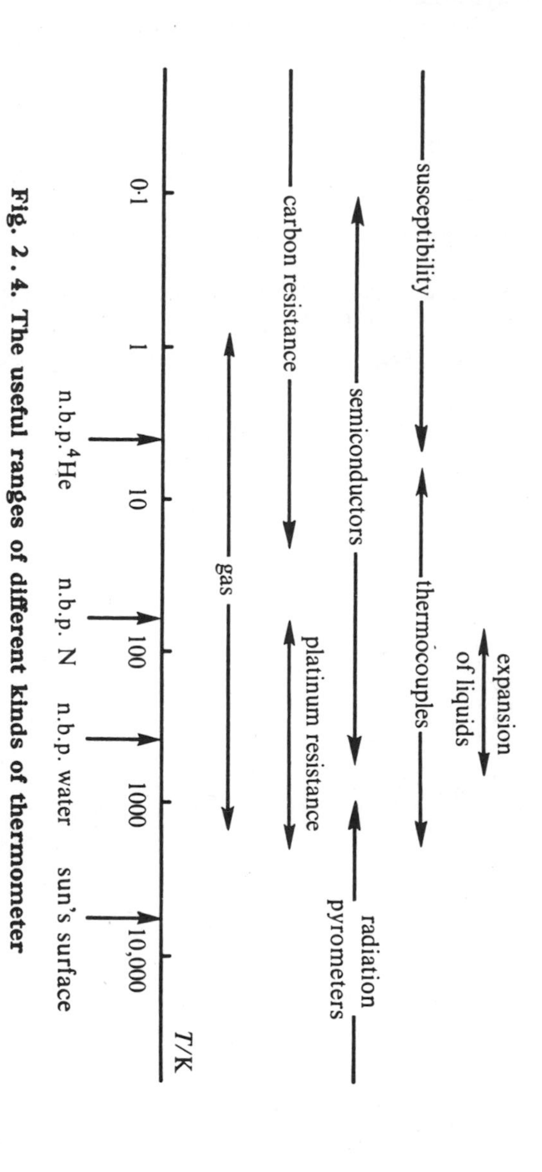

Fig. 2.4. The useful ranges of different kinds of thermometer

an *empirical* scale which is arranged to coincide as closely as possible with thermodynamic temperature.

The current scale[8] uses three thermometers:

Range	*Instrument*
13·81 K to 903·89 K	platinum resistance thermometer
903·89 K to 1337·58 K	platinum/platinum + 10% rhodium thermocouple
above 1337·58 K	radiation pyrometer

These are calibrated against the standard and primary reference points listed in Table 2.1. Reference 8 describes the 1968 International Practical Temperature Scale in detail, and also includes relevant experimental information.

[8] *The International Practical Temperature Scale of 1968*, Her Majesty's Stationery Office, London, 1969.

3. The First Law

3.1. The Background to the First Law

The first law of thermodynamics is essentially an extension of the principle of the conservation of energy to include systems in which there is flow of heat. Historically, it marks the recognition of heat as a form of energy.

The work which led up to this is well known. There were two rival theories of the nature of heat. According to the *caloric theory*, heat, or caloric, was an indestructible fluid which permeated matter and flowed from hot bodies to colder ones. According to the *molecular motion theory*, heat was associated with rapid vibrations of the molecules of which matter was composed. Of the two, the caloric theory had the greater support until the middle of the last century, although some of the most significant experiments were done much earlier.

In 1761 Black had studied the melting of ice. He noted that the temperature of a pail of ice-cold water placed in a warm room rose quite quickly, whereas, if the pail contained ice, the temperature remained constant for many hours while the ice melted. If caloric flowed into the pail from the surroundings when it contained ice-cold water, it must also do so when it contained ice. Therefore, he argued, ice-cold water must contain more caloric than ice. In 1799, Davy showed that both wax and ice could be made to melt by rubbing two pieces together. According to the caloric theory, rubbing squeezed caloric out of the solid so that the liquid produced by friction should contain *less* caloric than the solid. Clearly, the liquid could not at the same time contain both more and less heat than the solid.

At about the same time, Rumford showed that the heat produced by trying to drill a gun barrel with a blunt tool was apparently inexhaustible. It depended only on the continuance of work, and in no way was affected by the previous treatment that the metal had had. He argued that no material substance could be supplied indefinitely by a body, but that heat must be some form of motion imparted by the process of drilling.

However, it was not until Joule's work of the 1840's that the molecular motion theory was put on a sound basis by his demonstration of the direct quantitative equivalence of work and heat. In his experiments, he produced heating in various thermally isolated systems by performing work on them. He used many ways of doing the work: viscous dissipation in liquids, friction between solids, and, later, electrical heating. He compared the amounts of work required to produce a given amount of heat, using as his measure of heat the temperature rise which would be produced in unit mass of water.[1] He found that if the only effect of the work was to produce heating, then, in all cases, the amount of work and the corresponding amount of heat were in a fixed proportion to one another thus implying a direct equivalence of heat and work as forms of energy. These ideas are expressed more precisely through the formal statement of the first law and the development which follows from it.

3.2. The First Law

In his experiments, Joule compared heat and work as means of effecting a change of state. However, it is convenient to introduce the idea of heat as a form of energy by placing the emphasis rather differently. The first law does this by making a general statement about the behaviour of systems whose state is changed under conditions of thermal isolation. The formal statement is as follows:

If the state of an otherwise isolated system is changed by the performance of work, the amount of work needed depends solely on the change effected and not on the means by which the work is performed nor on the intermediate stages through which the system passes between its initial and final states.

The kind of experiment visualized in this statement is illustrated in Fig. 3.1, where different paths between the initial and final states are explored and the work required compared. In fact, such different paths have never been studied carefully. Once heat was accepted as a form of energy, the idea of the conservation of energy in this context was accepted readily enough. Certainly, the consequences of the first law have been tested thoroughly. If further justification is needed for it, it may be found in the truth of what follows from it.

[1] Making the reference to water required a knowledge of the relative thermal capacities of the materials involved. These had been found earlier by Black using the method of mixtures (see page 35), which he was the first person to develop as a calorimetric technique.

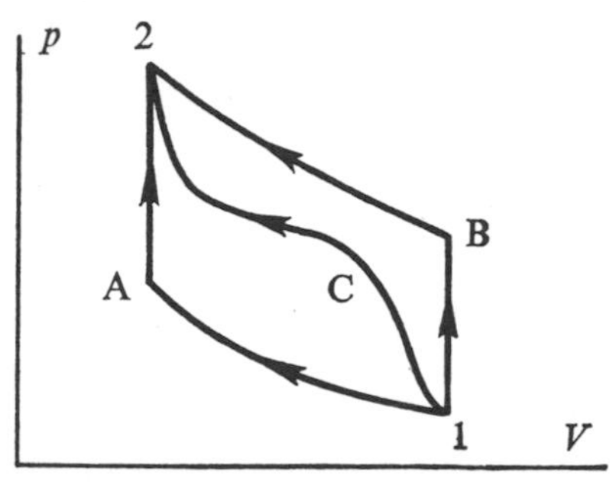

Fig. 3 . 1. Different adiabatic paths between two states of a fluid.

1A2: An adiabatic compression followed by electrical work at constant volume performed via a 'heater' of negligible thermal capacity immersed in the system.

1B2: The same processes, but in the reverse order.

1C2: A complex route requiring simultaneous electrical and mechanical work.

3.3. Internal Energy

If, as the first law states, a definite amount of energy is always associated with a given change of state (effected under adiabatic conditions), then the total energy of the system must be a function of state. We call it the *internal energy, U*. Thus, when a change of state is brought about by the performance of work alone, the work done *on* the system is simply the change in the internal energy in going from the initial to the final state:

$$\Delta U = W. \tag{3.1}$$

U is a function of state because W is independent of path.

3.4. Heat

Equation (3.1) applies to a thermally isolated system. However, we know that it is also possible to change the state of a system without doing work on it. We may use heat alone, or any combination of heat and work. Thus, when a system is not thermally isolated equation (3.1) is no longer valid. It must now be modified to

▶ $$\Delta U = Q + W \tag{3.2}$$

where Q is a measure of the extent to which the change is *not* adiabatic. Q is called *heat*. It is this form of the first law which has led to its common statement as 'Energy is conserved if heat is taken into account'.

We have thus defined heat as a form of energy entirely equivalent in its effect on the total energy of a system to energy communicated by the performance of some kind of work. The distinction between heat and work is not always clear-cut in the sense that it is not always easy to decide whether a particular energy contribution should be classed as heat or work. In the illustration of Fig. 3.1, we chose to count energy supplied by an electric 'heater' as work and this is entirely justified, for if the heater and fluid were placed together in an opaque box so that we knew nothing of the detailed composition of the system, we should certainly know only that a certain amount of electrical work (equal to $\int VI\,\mathrm{d}t$) was performed on it. Alternatively, we might have chosen not to make the heater part of the system but to attach it from outside by a thermal link. In this case, we should intuitively have considered the energy to be supplied as heat. Probably, the most convenient distinction is made in terms of whether the energy enters the system by a macroscopically ordered action or by one where order exists on the microscopic scale only. In the former case, the energy would be communicated by work and in the latter by heat. Thus, when a piston moves in a cylinder, the movement is macroscopic in the sense that the velocity of the piston is superimposed on all its molecules, and the *piston* does work on the gas. On the other hand, if the piston is hot, the (thermal) motions of its molecules are not correlated, energy is communicated to the gas by processes which are ordered on the microscopic scale only and we say that heat flows. That it should be impossible always to make a sharp distinction between heat and work is not surprising, for it is precisely the function of the first law to state that they are, in certain ways, equivalent.

According to the first law, then, when a system undergoes a given change, ΔU is necessarily defined since U is a function of state, and so the sum $Q+W$ is defined, but *not* Q *or* W *separately*. Only if we know *how* the system passes from its initial to its final state can Q and W be determined separately. Thus, for a given infinitesimal change we write

$$\mathrm{d}U = đQ + đW \tag{3.3}$$

where the symbol đ indicates that the infinitesimal quantities $đQ$ and $đW$ are not exact differentials: They cannot be evaluated from

[2] We have already had an example of a very simple constraint. For a thermally isolated system, $đQ = 0$. We shall later discuss more complicated constraints like $\mathrm{d}U = 0$.

a knowledge of the initial and final states alone; Q and W are not functions of state. However, if the system is constrained so that the path of the infinitesimal change is defined,[2] then $đQ$ and $đW$ are separately determined and may be treated as well behaved differentials. dU is, of course, always exact since U is a function of state. Equation (3.3) is known as the *differential form* of the first law.

In interactions between systems which are completely isolated from their surroundings, it is clear that the total internal energy must be conserved, for if we think of the group of systems as making up one larger complex system, no heat or work enters the composite system from outside so that its total energy cannot change. In the special case when the systems interact by exchange of heat alone, heat is also 'conserved', for we have

$$\Delta U_{\text{total}} = \sum \Delta U_i = \sum (Q_i + W_i) = 0.$$

But

$$W_i = 0$$

so

$$\sum Q_i = 0 \tag{3.4}$$

i.e., heat is conserved.

This is the basis of the *method of mixtures* used in calorimetry. Two or more systems are brought into thermal contact (mixed, perhaps literally), and in reaching thermal equilibrium, heat flows from one to another. Since no work is normally involved in the process (such effects as the change of hydrostatic pressure when a body is immersed are usually negligible) heat is conserved. If one substance, say water, is chosen as a reference, its temperature can be used as a measure of its internal energy, and its *change* in temperature is a measure of the heat exchanged. This was the principle used by Joule.

When using the first law in the form of equation (3.2) or (3.3), it is important to be clear about the signs of the terms. If ΔU is the change in internal energy of the system in going from its initial to its final state, then W must be the work done *on* the system and Q the heat transferred *to* the system. Having adopted this convention, it is possible to define *hotter* and *colder*. These are *comparative* terms and we define them by referring to the direction of heat flow when systems being compared are placed in thermal contact. We say that heat flows from the hotter body to the colder. Thus, in any given interval of time, Q is negative for the hotter body and positive for the colder.

That the first law in the form of (3.3) cannot, in general, be integrated is a disadvantage. Later, we shall replace $đQ$ and $đW$ by functions of state, and so obtain an equation in which all the terms are uniquely defined in any given change regardless of the path of the change. In this new form, the first law becomes much more powerful. However, we may only do this when we have considered the second law. For the moment we turn to the evaluation of W for various systems.

3.5. Work in Various Systems

It is always possible to express the work done on a system in terms of its parameters of state if the changes in which the work is performed are thermodynamically reversible. Only when this is the case do the system parameters describe the action of external forces. For example, if a fluid is enclosed in a cylinder by a piston with friction, then the force that has to be applied to the piston to overcome the friction and compress the fluid is greater than the force exerted on the fluid by the piston. Thus the external work done on the whole system (fluid and its container) is greater than the work done on the fluid alone and cannot be expressed in terms of the fluid's parameters of state. If the friction is negligible, however, the work done on the fluid becomes equal to that done by external forces and both can be expressed in terms of the system parameters.

When changes take place reversibly, the work done on a system is

$$đW = \sum_i X_i \, dx_i$$

where X_i and x_i are the forces and their conjugate displacements. We shall show that the work is of this form in several special cases.

3.5.1. Work by Hydrostatic Pressure

Consider a fluid contained in a cylinder by a frictionless, tight fitting piston (Fig. 3.2). Let the surface area of the piston be A and the pressure exerted by the fluid, p. Then the force which must be exerted on the piston to contain the fluid is $F = pA$.

Now, suppose the piston is moved *in* a small distance $d\xi$. The work done *on* the fluid is

$$đW = F \, d\xi = pA \, d\xi = -p \, dV$$

where dV is the change in volume of the fluid. The sign should be noted. In *increasing* its volume a fluid does work *on* its surroundings.

We shall use this result a great deal since, as is customary in following through the development of thermodynamics, we shall generally use a fluid system as a convenient model.

In the case of work done on a solid by hydrostatic pressure, we immerse the solid in an incompressible fluid and again contain the whole in a cylinder with piston (Fig. 3.3). Since the fluid is incompressible, no work can be done on it by changing the pressure and all the work done by the piston must be communicated to the solid. Thus for work done *on* a system by hydrostatic pressure we always have

$$đW = -p\,\mathrm{d}V. \tag{3.5}$$

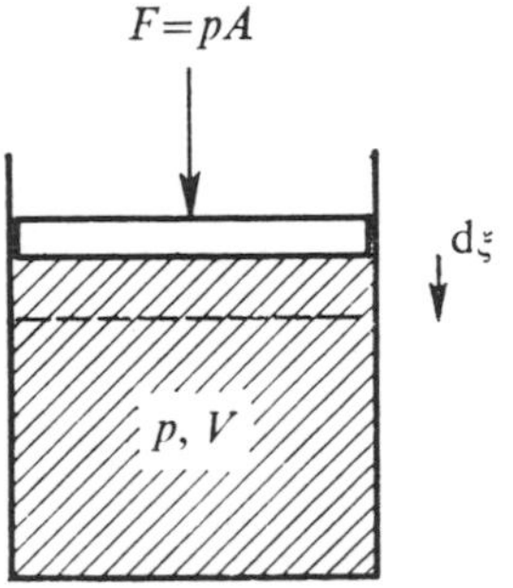

Fig. 3.2. Compression of a fluid.

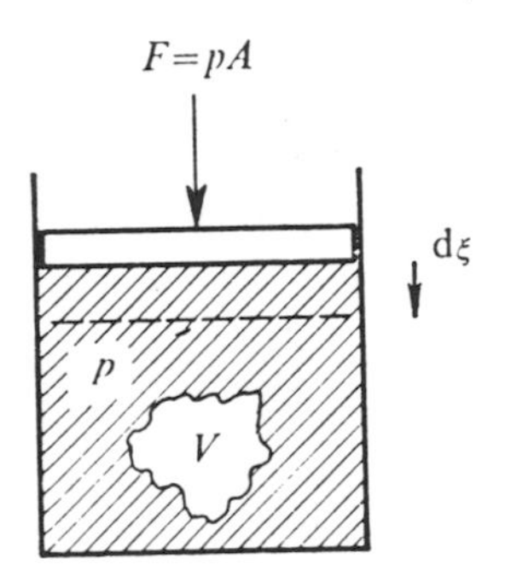

Fig. 3.3. Application of hydrostatic pressure to a solid.

To integrate (3.5) we need to know the conditions under which the change takes place. For an *isothermal* change in an ideal gas (see section 8.2), we may use the equation of state:

$$pV = RT$$

giving

$$W = p_1 V_1 \ln (p_2/p_1) = RT \ln (p_2/p_1). \tag{3.6}$$

For a reversible *adiabatic* change we have

$$pV^{\gamma} = \text{const.}$$

giving

$$W = \frac{1}{\gamma - 1}(p_2 V_2 - p_1 V_1). \tag{3.7}$$

In the case of a solid, we will calculate the work in the general case and then to simplify the result for the cases of simple constraints. (We could have done the same thing in the case of the gas.) We proceed as follows:

We must first make a choice of independent variables. Since we are often interested in changes occurring at constant pressure or constant temperatures we shall take p and T. $đW$ will then separate into two terms, one corresponding to pressure changes and one to temperature changes.

To put $đW = -p\,\mathrm{d}V$ in terms of the variables p and T we must substitute for $\mathrm{d}V$. In general, we will not know the functional form of V:

$$V = V(p, T).$$

Then

$$\mathrm{d}V = \left(\frac{\partial V}{\partial p}\right)_T \mathrm{d}p + \left(\frac{\partial V}{\partial T}\right)_p \mathrm{d}T$$

$$= -V\kappa_T \mathrm{d}p + V\beta_p \mathrm{d}T$$

where

$$\kappa_T = -\frac{1}{V}\left(\frac{\partial V}{\partial p}\right)_T = \text{isothermal compressibility},$$

and

$$\beta_p = \frac{1}{V}\left(\frac{\partial V}{\partial T}\right)_p = \text{isobaric cubic expansivity}.$$

Then

$$W = \int_{p_1}^{p_2} p\,\kappa_T V\,\mathrm{d}p - \int_{T_1}^{T_2} p\,\beta_p V\,\mathrm{d}T. \tag{3.8}$$

In the cases of isothermal and isobaric changes, we may simplify and approximate the general expression as follows. For an *isothermal* change we have

$$W = \int_{p_2}^{p_1} p\kappa_T V\,\mathrm{d}p \simeq \tfrac{1}{2}\kappa_T V(p_2^2 - p_1^2) \tag{3.9}$$

since V and κ_T are nearly constant for a solid. For an *isobaric* change,

$$W = -\int_{T_1}^{T_2} pV\beta_p\,\mathrm{d}T \simeq -pV\beta_p(T_2 - T_1) \tag{3.10}$$

since p is constant and V and β_p are nearly so for a solid.
Note that the work is again done by the hydrostatic pressure, but in this case the volume change is brought about by the expansion of the material caused by the temperature change.

3.5.2. Work Against Surface Tension

The surface tension, γ, of a liquid is defined as the work required isothermally to increase the area of surface by unity. Thus,

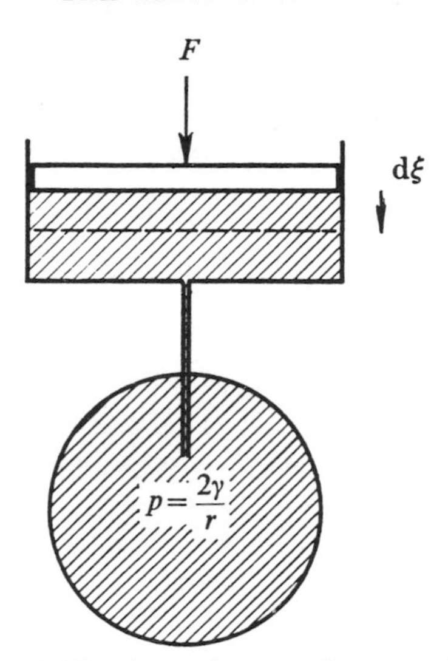

Fig. 3 . 4. Work against surface tension.

$$đW = \gamma \, dA. \tag{3.11}$$

It is found experimentally that γ is normally independent of area and depends only on temperature, so that for a finite increase of area under isothermal conditions we have

$$W = \gamma(A_2 - A_1).$$

It is a consequence of the fact that work is required to increase the area of a surface, that there is a pressure difference across a curved surface. The magnitude of the pressure difference may be derived by considering the work done as the surface is extended. We treat the spherical case.

Consider a drop of liquid suspended from the end of a fine capillary (Fig. 3.4). We increase the size of the drop by depressing the piston to force liquid down the capillary into the drop. If the excess pressure in the drop is p, the work done in a small displacement $d\xi$ of the piston is

$$đW = F \, d\xi = +p \, dV = p \, 4\pi r^2 \, dr$$

where $+dV$ is the volume change *of the drop* (and $-dV$ the volume change of the liquid in the cylinder; cf. equation (3.5)) and r is the drop radius. But, from the definition of surface tension, we also have

$$đW = \gamma \, dA = 8\pi\gamma r \, dr.$$

Equating the two expressions for $đW$ we obtain

$$p=\frac{2\gamma}{r}. \tag{3.12}$$

3.5.3. Work by an Electric Field

We use SI units defined by the equations[3]

$$\mathbf{D}=\epsilon_0\epsilon_r\mathbf{E}=\epsilon_0\mathbf{E}+\mathbf{P} \tag{3.13}$$

$$\chi_e=\epsilon_r-1=P/\epsilon_0 E \tag{3.14}$$

where **D** is electric displacement, ϵ_0 is the permittivity of a vacuum, ϵ_r is relative permittivity, **E** is electric field strength, **P** is electric polarization, and χ_e is electric susceptibility.

An electric field can do work on any polarizable material. Consider such a material, which need not be isotropic, to fill the space between the plates of a parallel-plate capacitor (Fig. 3.5). Let the separation of the plates be d and their area A, and suppose that $d^2 \ll A$ so that edge corrections may be neglected. When a potential difference is applied to the capacitor, an electric field is set up between the plates and the dielectric is polarized.

Let the potential difference across the capacitor be $\mathscr{E}$, and the total charge on the plates Z. Then, if **n** is a unit vector perpendicular to the plates, we have

$$\mathscr{E}=\mathbf{E}.\mathbf{n}d \tag{3.15}$$

and by Gauss' theorem,

$$Z=\mathbf{D}.\mathbf{n}A. \tag{3.16}$$

From the boundary condition on **E** at the surface of the conductors and symmetry, **E** must be parallel to **n** although, if the medium is not isotropic, **D** need not be.

If the charge on the capacitor is increased by a small amount dZ, the work done by the battery is

$$\begin{aligned} đW &= \mathscr{E}\,\mathrm{d}Z \\ &= Ad\,\mathbf{E}.\mathrm{d}\mathbf{D} \\ &= V\mathbf{E}.\mathrm{d}\mathbf{D} \end{aligned}$$

where V is the volume between the plates of the capacitor. With equation (3.13) this becomes

$$đW=\epsilon_0\mathbf{E}.\mathrm{d}\mathbf{E}\,V+\mathbf{E}.\mathrm{d}\mathbf{P}\,V. \tag{3.17}$$

[3] See B. I. Bleaney and B. Bleaney, *Electricity and Magnetism*, Chapter 1, Oxford University Press, 1965.

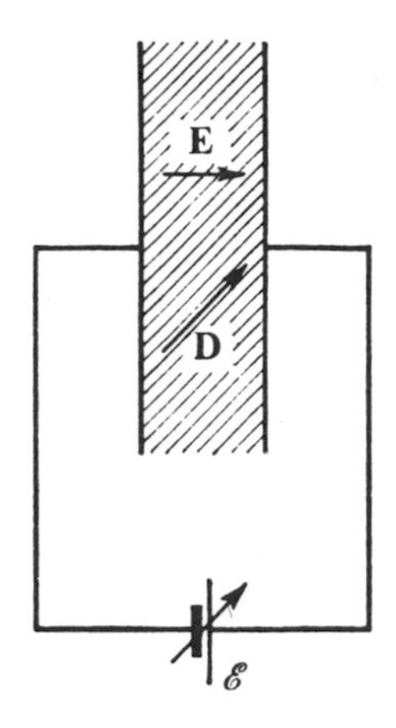

Fig. 3.5. Polarization of a dielectric.

Now this is the *total* work done on the volume subjected to the field. The first of the above terms is present in the absence of the material and represents the change of energy stored in the capacitor when empty. The work done on the dielectric is thus

$$đW = \mathbf{E}.\mathrm{d}\mathbf{P}\, V. \tag{3.18}$$

If the fields vary with position, this generalizes to

$$đW = \int (\mathbf{E}.\mathrm{d}\mathbf{P})\, \mathrm{d}V, \tag{3.19}$$

and if **E** and **P** are uniform throughout the material, equation (3.19) becomes

$$đW = \mathbf{E}.\mathrm{d}\mathbf{p} \tag{3.20}$$

where **p** is the total electric dipole moment of the specimen.

3.5.4. Work by a Magnetic Field

We use SI units defined by the equations[4]

$$\mathbf{B} = \mu_0\mu_r\mathbf{H} = \mu_0(\mathbf{H}+\mathbf{M}) \tag{3.21}$$

$$\chi_m = \mu_r - 1 = M/H \tag{3.22}$$

where **B** is magnetic induction, μ_0 is the permeability of a vacuum, μ_r is relative permeability, **H** is magnetic field strength, **M** is magnetization, and χ_m is magnetic susceptibility.

[4] See B. I. Bleaney and B. Bleaney, *Electricity and Magnetism*, Chapter 5, Oxford University Press, 1965.

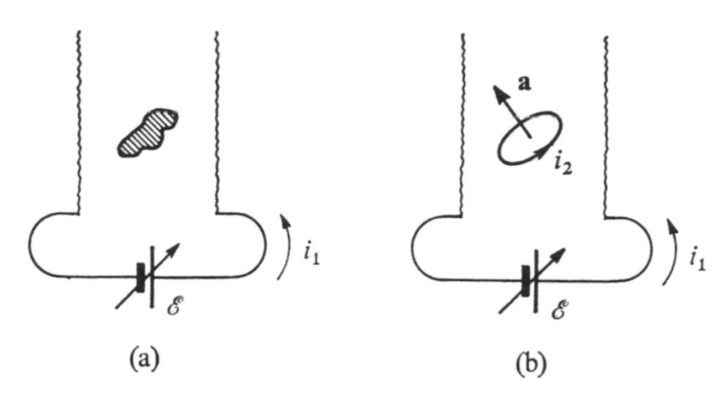

Fig. 3 . 6. Calculating the work done by a magnetic field.

A magnetic field can do work on any magnetizable material. Consider such a material to be subjected to a magnetic field by being placed inside a solenoid (Fig. 3.6a). Without loss of generality we may assume the solenoid to be resistanceless, so that the battery only does work against e.m.f.'s which are induced in the solenoid either as a direct result of changing the current (via the solenoid's self-inductance L_{11}) or by changes in the magnetization of the material which will also alter the flux linking the electric circuit. For convenience we represent an elementary dipole in the material by a small current loop of area **a** carrying a current i_2 (Fig. 3.6b). We need not assume that the magnetic induction in the solenoid is uniform. Instead, we put $\mathbf{B}=\mathbf{b}i$, where i is the current in the solenoid and **b** may vary with position.

The magnetic moment of the elementary current loop is

$$\mathbf{m}'=i_2\mathbf{a} \tag{3.23}$$

and the mutual inductance between the loop and the solenoid is

$$L_{12}=\mathbf{b}\,.\,\mathbf{a}. \tag{3.24}$$

Hence, the back e.m.f. in the solenoid is

$$L_{11}\frac{\mathrm{d}i_1}{\mathrm{d}t}+\sum_2 L_{12}\frac{\mathrm{d}i_2}{\mathrm{d}t}$$

where the summation is over all the elementary current loops. The rate of working by the battery is therefore

$$\frac{\text{đ}W}{\mathrm{d}t}=i_1\left[L_{11}\frac{\mathrm{d}i_2}{\mathrm{d}t}+\sum_2 L_{12}\frac{\mathrm{d}i_2}{\mathrm{d}t}\right].$$

Hence, during an infinitesimal change in conditions, the work done is

$$đW = i_1 L_{11}\, \mathrm{d}i_1 + i_1 \sum_2 L_{12}\, \mathrm{d}i_2$$

$$= i_1 L_{11} \mathrm{d}i_1 + \sum \mathbf{B}.\mathrm{d}\mathbf{m}' \tag{3.25}$$

where we have used equations (3.23) and (3.24). The first term, however, is the work that would have been done in the absence of the magnetic material and is the change in the energy stored in the inductor when empty. The work done on the material is thus

$$đW = \sum \mathbf{B}.\mathrm{d}\mathbf{m}'.$$

If the fields vary with position, this generalizes to

$$đW = \int (\mathbf{B}.\mathrm{d}\mathbf{M})\, \mathrm{d}V, \tag{3.26}$$

and if **B** and **M** are uniform throughout the material, equation (3.26) becomes

$$đW = \mathbf{B}.\mathrm{d}\mathbf{m} \tag{3.27}$$

where **m** is the *total* magnetic moment of the specimen.[5]

Some of the various expressions for the work done on a system in an infinitesimal reversible change are collected together in Table 3.1. We have not given a formal derivation of the first two since they follow trivially from fundamental definitions. It should be pointed out that when writing down the first law in the form $\mathrm{d}U = đQ + đW$, all the effective work terms must be included in $đW$.

3.6. Heat Capacities

The first law allows us to express the heat absorbed by a system during a reversible change as

$$đQ = \mathrm{d}U - \sum X_i \mathrm{d}x_i.$$

$đQ/\mathrm{d}\alpha$ is called a *general heat capacity* of the system and expresses the rate at which heat is absorbed when the variable α is changed. α may be any function of state. As usual, the heat absorbed in a given change is not defined unless the path of the change is defined.

[5] We shall use this form for magnetic work throughout this book. However, it should be noted that it is possible to specify the thermodynamic system in a different way so as to obtain an alternative expression which is more useful in the context of statistical mechanics. The difference between the two forms is discussed in the appendix.

Table 3.1. A Summary of Some Common Contributions to the Work Done on a System in an Infinitesimal Reversible Change

System subject to work by:	$đW$
tensional force	$\mathscr{F}\,\mathrm{d}L$
electric current*	$\mathscr{E}\,\mathrm{d}Z$
hydrostatic pressure	$-p\,\mathrm{d}V$
change of surface area	$\gamma\,\mathrm{d}A$
electric field	$\mathbf{E}.\mathrm{d}\mathbf{p}$
magnetic field	$\mathbf{B}.\mathrm{d}\mathbf{m}$
generalized forces	$X_i\,\mathrm{d}x_i$

* $\mathscr{E}$ is potential difference and Z is charge.

Hence $đQ/\mathrm{d}\alpha$ remains undefined in the absence of constraints. To define the path of a change in an n-parameter system, we need $(n-1)$ constraints. The heat capacity is then defined and it may be written as

$$\frac{đQ_{\beta,\,\gamma,\,\ldots}}{\mathrm{d}\alpha}$$

where $\beta, \gamma, \ldots$ are the constraints.[6]

The various heat capacities areusually represented by the symbol C to which is added a suffix indicating the constraints and a superscript in brackets showing the variable with respect to which the differential is made. Thus, we write

$$\frac{đQ_{\beta,\,\gamma,\,\ldots}}{\mathrm{d}\alpha}=C^{(\alpha)}_{\beta,\,\gamma\,\ldots}.$$

For example,

$$\frac{đQ_V}{\mathrm{d}T}=C^{(T)}_V$$

is the rate at which heat is absorbed as the Kelvin temperature is changed with the volume kept constant.

In the case of thermal capacities for increase of temperature, the superscript is usually omitted, as with the principal[7] heat capaci-

[6] Differentials of this kind are sometimes called *curve differentials* since the possible states of the constrained system may be represented by a curve in system-coordinate space. For example, the possible states of a gas held at a constant pressure are represented by an isobar. $đQ_p/\mathrm{d}T$ measures the rate at which heat is absorbed as the system is caused to move along the isobar by increasing its temperature.

[7] See section 8.1.

ties, C_p and C_V.

Heat capacities are clearly extensive quantitities and it is often convenient to use the heat capacity per unit mass: the specific heat capacity. This is usually called simply the *specific heat*, and, as usual, is given a small letter symbol. Thus, c_p and c_V are the principal specific heats.

We may illustrate these general remarks by reference to a simple system subject to work by hydrostatic pressure. The first law becomes

$$đQ = \mathrm{d}U + p\,\mathrm{d}V.$$

Suppose we wish to use T and V as independent variables. Then we must express $\mathrm{d}U$ in terms of $\mathrm{d}T$ and $\mathrm{d}V$:

$$\mathrm{d}U = \left(\frac{\partial U}{\partial T}\right)_V \mathrm{d}T + \left(\frac{\partial U}{\partial V}\right)_T \mathrm{d}V.$$

Substituting,

$$đQ = \left(\frac{\partial U}{\partial T}\right)_V \mathrm{d}T + \left[p + \left(\frac{\partial U}{\partial V}\right)_T\right] \mathrm{d}V.$$

If we apply the constraint that V is constant, we have the usual thermal capacity at constant volume,

$$C_V^{(T)} = \frac{đQ_V}{\mathrm{d}T} = \left(\frac{\partial U}{\partial T}\right)_V. \qquad (3.28)$$

(This is an obvious result since if there is no volume change no work can be done and any change in internal energy must simply be equal to the heat entering the system.) The corresponding specific heat is

$$c_V = \frac{1}{m}\left(\frac{\partial U}{\partial T}\right)_V = \left(\frac{\partial u}{\partial T}\right)_v.$$

However, if T is constant, we have a new kind of thermal capacity:

$$C_T^{(V)} = \frac{đQ_T}{\mathrm{d}V} = p + \left(\frac{\partial U}{\partial V}\right)_T \qquad (3.29)$$

being the amount of heat absorbed per unit volume increase as the system moves *along an isotherm*: a sort of latent heat.

We may obtain similar expressions if we choose p and T as our independent variables. To substitute in the first law we first obtain $\mathrm{d}U$ and $\mathrm{d}V$ in terms of $\mathrm{d}p$ and $\mathrm{d}T$:

$$\mathrm{d}U = \left(\frac{\partial U}{\partial p}\right)_T \mathrm{d}p + \left(\frac{\partial U}{\partial T}\right)_p \mathrm{d}T$$

and

$$\mathrm{d}V=\left(\frac{\partial V}{\partial p}\right)_T \mathrm{d}p+\left(\frac{\partial V}{\partial T}\right)_p \mathrm{d}T.$$

Then

$$đQ=\left[\left(\frac{\partial U}{\partial p}\right)_T+p\left(\frac{\partial V}{\partial p}\right)_T\right]\mathrm{d}p+\left[\left(\frac{\partial U}{\partial T}\right)_p+p\left(\frac{\partial V}{\partial T}\right)_p\right]\mathrm{d}T. \quad (3.30)$$

In this case, the two heat capacities are

$$C_T^{(p)}=\left(\frac{\partial U}{\partial p}\right)_T+p\left(\frac{\partial V}{\partial p}\right)_T \quad (3.31)$$

and

$$C_p^{(T)}=\left(\frac{\partial U}{\partial T}\right)_p+p\left(\frac{\partial V}{\partial T}\right)_p. \quad (3.32)$$

Similar expressions are obtained with p and V as independent variables.

It should be noted that substitution of (3.31) and (3.32) in (3.30) gives

$$đQ=C_T^{(p)}\,\mathrm{d}p+C_p^{(T)}\,\mathrm{d}T$$

which is simply an expansion of $đQ$ in terms of $\mathrm{d}p$ and $\mathrm{d}T$:

$$đQ=\frac{đQ_T}{\mathrm{d}p}\,\mathrm{d}p+\frac{đQ_p}{\mathrm{d}T}\,\mathrm{d}T,$$

and looks like the usual partial differential form; but it must be remembered that Q is *not* a function of p and T, and that although all the terms on the right-hand side of the equation are well behaved functions of state, their integrals still depend on path and $đQ$ is undefined in the absence of a constraint.

3.7. Enthalpy

C_V took a particularly simple form in terms of the internal energy:

$$C_V=\left(\frac{\partial U}{\partial T}\right)_V \quad \text{(equation (3.28))}$$

but the other common heat capacity had a more complicated form:

$$C_p=\left(\frac{\partial U}{\partial T}\right)_p+p\left(\frac{\partial V}{\partial T}\right)_p. \quad \text{(equation (3.32))}$$

It would be convenient to construct an energy function H, which would give C_p in a form similar to (3.28). That is,

$$C_p = \left(\frac{\partial H}{\partial T}\right)_p. \tag{3.33}$$

Equating (3.33) and (3.32),

$$\left(\frac{\partial H}{\partial T}\right)_p = \left(\frac{\partial U}{\partial T}\right)_p + p\left(\frac{\partial V}{\partial T}\right)_p = \left[\frac{\partial}{\partial T}(U+pV)\right]_p.$$

The function we require is thus

▶ $$H = U + pV. \tag{3.34}$$

H is called *enthalpy*. As all the terms on the right-hand side of (3.34) are functions of state, H must also be a function of state. Its differential form follows immediately from (3.34):

$$dH = dU + p\,dV + V\,dp = đQ + V\,dp. \tag{3.35}$$

We see from (3.33) and (3.35) that when a system undergoes a reversible isobaric change, the change in H is equal to the heat absorbed:

$$dH = đQ_p. \tag{3.36}$$

Since many processes of interest do take place at constant pressure, H has sometimes been given the misleading name of *heat content*.

Equation (3.36) is true whatever the system. In particular, it also applies to one in which a chemical reaction takes place. In this case, ΔH becomes the heat of reaction. This property of enthalpy makes it an important quantity in chemical thermodynamics.

3.8. Flow Processes

Any process in which there is a steady flow of some *working substance* through a device which permits the transformation of internal energy into external energy (e.g., potential energy) or work is called a *flow process*. Enthalpy is an important quantity for such processes.

The general case of a flow process is represented in Fig. 3.7. Let the parameters of the working substance at the inlet and outlet be

pressure	p_1 and p_2
specific volume	v_1 and v_2
specific internal energy	u_1 and u_2
speed of flow	$\mathscr{V}_1$ and $\mathscr{V}_2$
height	z_1 and z_2.

Suppose also that in flowing through the device *unit mass* does external work w (work *on* the surroundings *by* the working substance), and *absorbs* heat q. Then the energy transported into the device by *unit mass* is

$$u_1+\tfrac{1}{2}\mathscr{V}_1^2+gz_1.$$

(Internal energy plus kinetic energy plus potential energy.)

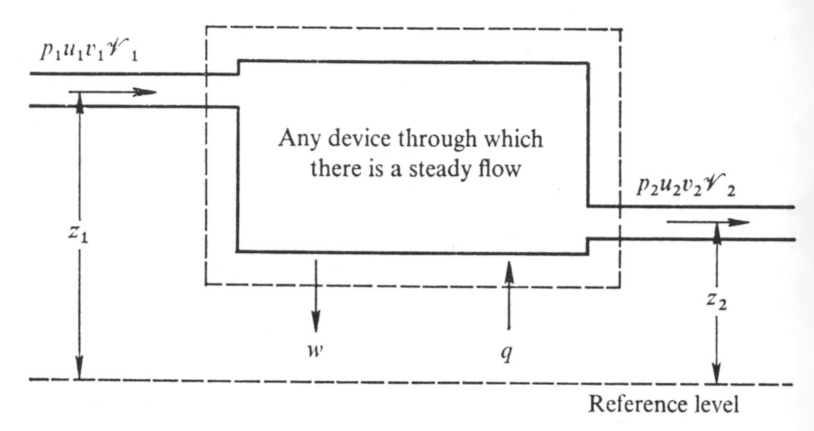

Fig. 3 . 7. The general flow process.

However, this is not the total energy entering the device at the inlet, for the source of the working substance also does mechanical work on the device as the fluid enters. (One may visualize that the pressure is maintained by a piston which has to move forward against the pressure p_1 to force the fluid in.) As unit mass flows, the work done is

$$\int p\,\mathrm{d}V=p_1v_1.$$

This contribution is known as *flow work*. Similarly, at the outlet the device has to supply work to force the working substance out against the pressure p_2.

Thus, the total energy transferred across the inlet as unit mass flows is

$$u_1+p_1v_1+\tfrac{1}{2}\mathscr{V}^2+gz_1,$$

and a similar expression applies at the outlet. Then, since conditions are steady, we may apply the conservation of energy to the region of space enclosed by the dotted line in Fig. 3.7. This gives

$$(u_1+p_1v_1+\tfrac{1}{2}\mathscr{V}_1^2+gz_1)-(u_2+p_2v_2+\tfrac{1}{2}\mathscr{V}_2^2+gz_2)=w-q$$

i.e.,

$$w=(h_1-h_2)+\tfrac{1}{2}(\mathscr{V}_1^2-\mathscr{V}_2^2)+g(z_1-z_2)+q \qquad (3.37)$$

where h is the specific enthalpy.

Equation (3.37) is essentially the general form of the first law applied to steady mass flow. To calculate *how* the enthalpy changes in a particular flow process we require the results of the second law; but it should be noted that, thus far, we have had to make no assumptions about the nature of the processes taking place in the device. In particular, they need not be thermodynamically reversible. Thus (3.37) may be applied to a wide range of processes and is of considerable importance in many branches of pure and applied science. A few simple cases will serve to illustrate this.

3.8.1. The Constant Flow Calorimeter

In a constant flow calorimeter, kinetic and potential energy terms are usually negligible and no work is exchanged with the surroundings. In this case, (3.37) reduces to

$$h_2 - h_1 = q \tag{3.38}$$

i.e., the heat absorbed is equal to the increase in enthalpy.

If the fluid is a liquid, energy changes due to (thermal) expansion are normally negligible and $(h_2 - h_1) = c_p \Delta T \simeq c_V \Delta T$. In the case of a gas, however, expansion has to be taken into account and the application of (3.38) is less trivial.

3.8.2. The Porous Plug or Throttle Valve

The purpose of a porous plug or throttling valve is to reduce the pressure of a flowing fluid without doing external work. Since heat exchanges and work against gravity are usually unimportant, (3.37) reduces to

$$h_1 - h_2 = \tfrac{1}{2}(\mathscr{V}_2^2 - \mathscr{V}_1^2). \tag{3.39}$$

In many cases, the kinetic energy before expansion (i.e., on the high pressure side) is also negligible and we simply have

$$h_1 - h_2 = \tfrac{1}{2}\mathscr{V}_2^2. \tag{3.40}$$

The fluid achieves a large drop in pressure and a small amount of kinetic energy is produced. In the limiting case, where there is sufficient friction to make all kinetic energy terms negligible we simply have $h_1 = h_2$. This is the condition which applies in the Joule–Kelvin expansion which will be dealt with more fully after we have discussed the second law. It should be noted that a process of this sort is highly irreversible.

3.8.3. The Ideal Nozzle

The ideal nozzle is the opposite extreme from the perfect throttle. Here, the intention is to create as high a velocity as possible by keeping friction and turbulence small. Often, the kinetic energy before the nozzle is small and we again have

$$h_1 - h_2 = \tfrac{1}{2}\mathscr{V}_2^2 \qquad \text{(equation (3.40))}$$

but now there is a large drop in enthalpy and a high kinetic energy after expansion.

In the jet engine the w and z terms of (3.37) are negligible. The fuel provides a large q and so raises the specific enthalpy of the gas which has been drawn into the system to a high value. This is accompanied by a large rise in pressure. The high enthalpy fluid is then expanded in the nozzle at the back of the engine, in the course of which the enthalpy is reduced and the energy converted into kinetic form in the emerging gases. The enthalpy is never reduced to its initial low value but the whole process tends towards a complete conversion of q into kinetic energy. The purpose of after-burners on a jet engine is to increase $\mathscr{V}_2$ still more by adding yet more heat after the initial expansion.

3.8.4. The Turbine

In the turbine, the object is to obtain the greatest possible external work. In this case, the device is designed to reduce kinetic energy terms to a minimum and (3.37) becomes

$$h_1 - h_2 = w. \qquad (3.35)$$

3.8.5. Streamline Flow

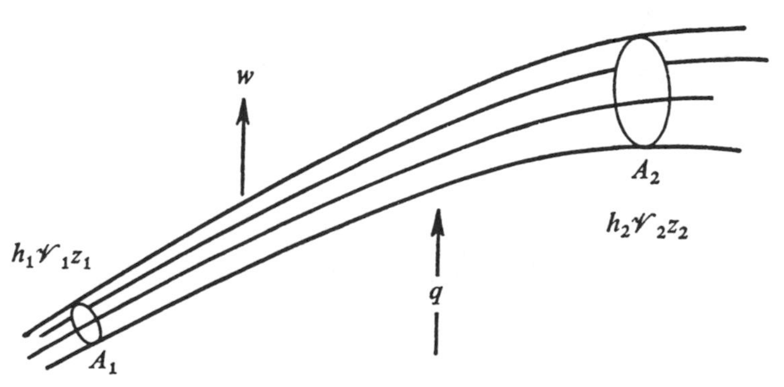

Fig. 3.8. Application of the first law to streamline flow.

Consider an area A_1 in a region of steady streamline flow (Fig. 3.8). The streamlines which pass through the edges of A_1 generate a tube of flow. Equation (3.37) may be applied to any length of such a tube. w becomes the viscous work, if any, across the bounding streamlines and q the heat entering across them.

When viscous losses and flow of heat are negligible, (3.37) simplifies to

$$u + p/\rho + \tfrac{1}{2}\mathscr{V}^2 + \phi = \text{const} \tag{3.42}$$

where ρ is the density and ϕ the potential energy per unit mass. The constancy of this quantity along a streamline is a fundamental result of hydrodynamics known as *Bernoulli's theorem*. If the fluid is incompressible, this further simplifies to

$$p/\rho + \tfrac{1}{2}\mathscr{V}^2 + \phi = \text{const.} \tag{3.43}$$

4. The Second Law

4.1. The Function of the Second Law

The first law of thermodynamics is a generalization of the principle of the conservation of energy to include heat. It places a restriction on the changes of a system which are *energetically* possible. Not all such changes occur however, and we have already acknowledged this fact in discussing thermal equilibrium and hotness. If two bodies are placed in thermal contact it would be energetically possible for their temperatures to diverge; it would not violate the first law. However, we know that this does not happen. The temperatures converge and eventually thermal equilibrium is established. Thus there is an essential *irreversibility of nature*, a natural direction for change, which we need to take into account in trying to describe thermal processes. The first function of the second law is to express this irreversibility.

Secondly, although we know that work may be converted into heat by a suitable dissipative mechanism (Joule's paddle wheels, or a resistor), we have not examined the conversion of heat into work. The first law emphasized the equivalence of heat and work as forms of energy, but it tells us nothing about the conversion from one form to the other; and, in particular, it tells us nothing about the *efficiency* with which heat may be converted into work, a matter of enormous practical importance. The second function of the second law is to express the inherent limit to the efficiency with which heat may be converted into work.

4.2. Cyclic Processes and Heat Engines

Now, in order to convert heat into work, we require some suitable thermodynamic machine which will consume heat and produce work. Clearly, if we are to discuss the efficiency of conversion, the machine itself must not suffer any permanent change in use. It must play a passive role in the sense that after it has completed an appropriate series of processes it must return to its initial state. Any series of processes by which a system is returned to its initial state is called a *cycle*. The machine, which will contain the system and a mechanism for causing it to undergo the cycle, is then called a *heat engine*, and it is convenient to refer to the system with which the engine operates as the *working substance.*

In general, a heat engine will absorb and reject heat during various parts of its cycle. We must expect there to be some heat rejected, for else (applying the first law to the engine) all the absorbed heat would have to be converted into work, giving 100 per cent efficiency which, as we shall see, is unobtainable even with an ideal heat engine.

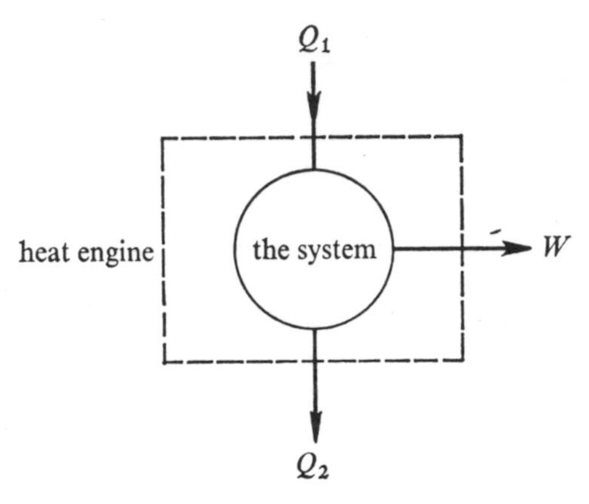

Fig. 4 . 1. Application of the first law to a heat engine.

We define the *thermal efficiency* η of a heat engine as the proportion of the absorbed heat which is turned into work:

$$\eta = \frac{\text{Work out}}{\text{Heat in}} = \frac{W}{Q_1},$$

where Q_1 and W are the heat absorbed and the work done in one cycle. Then, applying the first law to one cycle of the heat engine in the form $W = Q_1 - Q_2$, where Q_2 is the heat rejected, we have (see Fig. 4.1),

$$\eta = 1 - \frac{Q_2}{Q_1}. \tag{4.1}$$

There is one particularly simple cyclic process which plays an important part in the development of thermodynamics. Known as a *Carnot cycle*, it consists of four distinct processes:

(a) The working substance expands isothermally and reversibly at temperature Θ_1 absorbing heat Q_1.

(b) The working substance expands adiabatically and reversibly, the temperature changing from Θ_1 to Θ_2.
(c) The working substance is compressed isothermally and reversibly at Θ_2 rejecting heat Q_2.
(d) The working substance is compressed adiabatically and reversibly from Θ_2 to the initial state at Θ_1.

The cycle thus consists of the intersection of two adiabatics and two isothermals. During each part of the cycle, work is exchanged with the surroundings so that the net work done *by* the system in the whole cycle is

$$W = -\sum \oint X_i \, dx_i.$$

For a simple two-parameter system, this is numerically equal to the area enclosed by the cycle when plotted in the X–x coordinate plane.

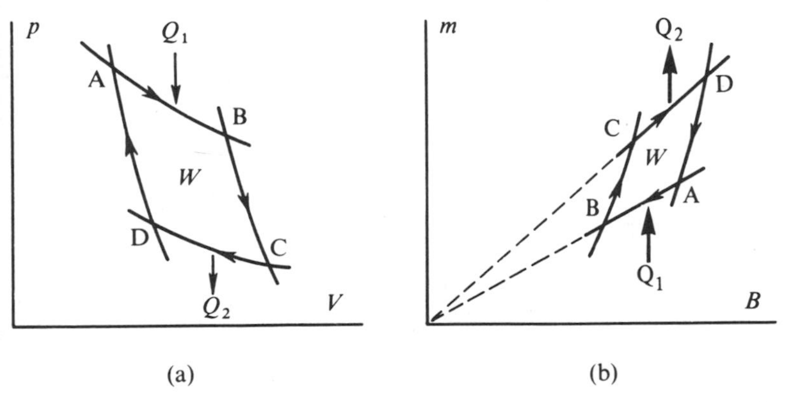

Fig. 4.2. Carnot cycles in a gas, (*a*), and a paramagnetic material, (*b*). AB and CD are isothermal processes. BC and DA are adiabatic.

Carnot cycles for two simple systems are illustrated in Fig. 4.2. The two examples chosen show how dissimilar Carnot cycles for different systems may appear. In the case of an ideal gas, the isotherms are rectangular hyperbolae, while for a paramagnetic material obeying Curie's law, $\chi = a/T$, they are straight lines through the origin.

Since every process in a Carnot cycle is reversible, the whole cycle must be reversible. Driven backwards, we shall see that a Carnot engine extracts heat from a body at a colder temperature and rejects heat to one at a hotter: it transfers heat in the 'unnatural' direction at the expense of mechanical work.

It should be pointed out that if a single working substance undergoes a reversible cycle in which heat is exchanged at two temperatures only, then the cycle is necessarily a Carnot cycle, for those parts of the cycle not involving the transfer of heat must necessarily be (reversible) adiabatics.

4.3. The Statements of the Second Law

There are two well-known statements of the second law.[1] The first places the emphasis on the *efficiency* of conversion of heat into work and the second on the irreversibility of nature.

The Kelvin Statement:

No process is possible whose sole result is the complete conversion of heat into work.

The Clausius Statement:

No process is possible whose sole result is the transfer of heat from a colder to a hotter body.

The Kelvin form states that it is impossible to achieve 100 per cent efficiency in the conversion of heat into work. In the Clausius form, the law denies the possibility of reversing the natural tendency for heat to flow from hotter to colder *without external interference* (in the form of work, for example).

We may simply demonstrate these two statements of the second law to be equivalent by showing that if one is untrue, then the other must be untrue also. We shall prove this one way round only. We shall show that if the Clausius statement is untrue then the Kelvin one must also be untrue. To do this we start with an engine which violates the Clausius statement. We then combine this engine with a normal heat engine (i.e., one which does not violate either statement) in such a way as to construct a composite engine which violates the Kelvin statement. We proceed as follows:

Suppose we have an engine which violates the Clausius statement of the second law by extracting heat from a cold reservoir at Θ_2

[1] There is also a third statement due to Carathéodory which we shall discuss in chapter 6. Although his formulation is much more economical than those of Kelvin or Clausius, it is framed in less practical terms, and its approach will be appreciated more easily when we have developed the subject a little further.

and delivering heat to a hotter reservoir at Θ_1. (Engine 1 of Fig. 4.3.) Since no work is involved the heat absorbed must equal that delivered in each cycle. We now take *any* heat engine and operate it in the normal way so that heat is absorbed at the hotter reservoir, rejected at the colder and work is done. (Engine 2 of Fig. 4.3.) Suppose that in some given time the first engine transfers Q_2 from Θ_2 to Θ_1. Then we operate the second engine at such a rate that during the same period it rejects Q_2 at Θ_2. If the heat it absorbs in that period is Q_1, the work done is $W = Q_1 - Q_2$. We now consider the two engines, taken together, to form a composite engine, and

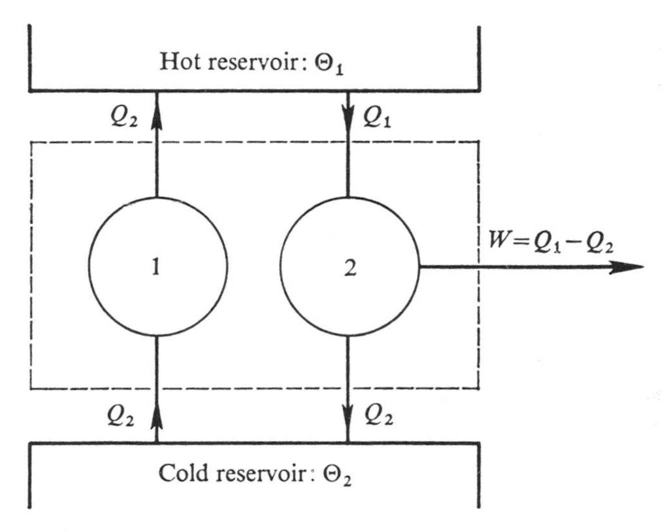

Fig. 4.3. The proof that the untruth of the Clausius statement of the second law implies the untruth of the Kelvin statement.

we see that the net effect of the composite engine is to exchange no heat at the cold reservoir but only to extract heat, $(Q_1 - Q_2)$, at Θ_1 and do an equal amount of work. The existence of such an engine violates the Kelvin statement.

The proof that if the Kelvin statement is untrue then the Clausius statement is also, proceeds in an analogous manner. Both proofs together show that the truth of either form of the second law is both a necessary and a sufficient condition for the truth of the other.

It is worth pointing out that the first and second laws imply the impossibility of two different forms of perpetual motion. The first law does not allow *perpetual motion of the first kind:* a machine cannot operate continuously by creating its own energy (because energy is conserved). The second law forbids *perpetual motion of the second kind:* a machine cannot be made which runs continuously by using the internal energy of a single heat reservoir. (Kelvin statement.) This would not violate the first law. A further possible way of obtaining perpetual motion would be to remove all dissipative effects such as friction, viscosity, or electrical resistance, so that motion, once started in some device, would persist. This would be *perpetual motion of the third kind*. It is not forbidden by either the first or the second law but it is known from experience to be impossible to achieve in any system governed by classical laws.[2]

4.4. Hotness and Temperature

We may now pause to clear up a minor point which we have glossed over somewhat. We have defined temperature in terms of thermal *equilibrium* between two bodies. We have defined hotness in terms of the natural direction of heat flow between bodies which are *not* in equilibrium. But we have *assumed* that there is a direct correspondence between hotness and temperature. This assumption is well founded on experience but, in fact, it is possible to prove its correctness directly from the second law. It will suffice for this proof if we may show the truth of the following statement: If one body at Θ_1 is hotter than one body at Θ_2, all bodies at Θ_1 are hotter than all bodies at Θ_2.

To prove this theorem we make use of a special kind of heat leak which consists of a Carnot engine C whose work output is dissipated as heat at the colder reservoir. It thus serves to transfer heat in the natural direction, from hot to cold, but by a process we may analyse.

Consider two of the bodies at Θ_1, A and B, and one at Θ_2, D, and suppose that A is hotter than D (Fig 4.4). We first operate the Carnot heat leak between A and D. Then, by the Clausius statement of the second law, the direction of heat flow must be from A to D. The isothermal process which the Carnot engine performs while in contact with A is determined by the condition that the working substance in the engine be in thermal equilibrium with A throughout the process. But, from the definition of empirical temperature, B is

[2] Superconductivity, the lossless flow of electric current, and superfluidity, the viscousless flow of one form of liquid helium, are essentially macroscopic quantum effects. These will be discussed in more detail in sections 10.9.3 and 10.9.2.

in equilibrium with A, and therefore, by the zeroth law, the Carnot engine would also be in equilibrium with B throughout the Θ_1 isothermal process. Thus, the Carnot engine would perform identical cycles whichever body at Θ_1 provides the heat source, and consequently the direction of heat flow will be the same. It therefore follows that, unless the Clausius statement is to be violated, any body at Θ_1 must be hotter than D. By applying a similar argument to bring in other bodies at Θ_2, this conclusion immediately generalizes into the statement which we originally set out to prove. Thus the unique correspondence of hotness and temperature is established.

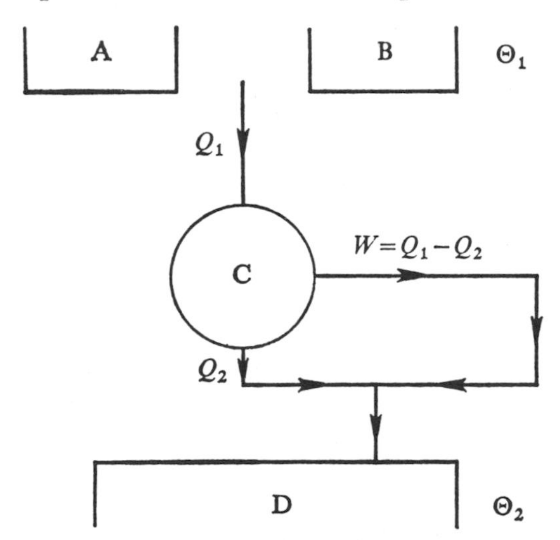

Fig. 4.4. Demonstration of the correspondence between hotness and temperature.

4.5. Carnot's Theorem

Carnot's theorem is the first stage in the derivation of the absolute temperature scale. The theorem states: *No engine operating between two given reservoirs can be more efficient than a Carnot engine operating between the same two reservoirs.* To prove this we show that if the theorem is untrue, we may construct out of a Carnot engine and one of these more efficient engines a composite engine which violates one of the statements of the second law. We shall arrange to violate the Clausius statement.

Consider a Carnot engine C and the hypothetical engine of greater efficiency, H, operating between reservoirs at Θ_1 and Θ_2. The energy changes during one cycle of the engines are shown in Fig. 4.5. If the efficiency of the hypothetical engine is greater than that of the Carnot engine, we have

$$\eta_H > \eta_C$$

i.e.,

$$\frac{W_H}{Q_{H1}} > \frac{W_C}{Q_{C1}}. \tag{4.2}$$

Since a Carnot engine is reversible we may drive it backwards with the mechanical energy from H. We may also choose the size of the Carnot engine's cycle, for although the isotherms are fixed by Θ_1 and Θ_2 we may move the position of the adiabatics as we please. This allows us to arrange that, in one cycle of each engine, C uses exactly as much mechanical work as H produces.[3]

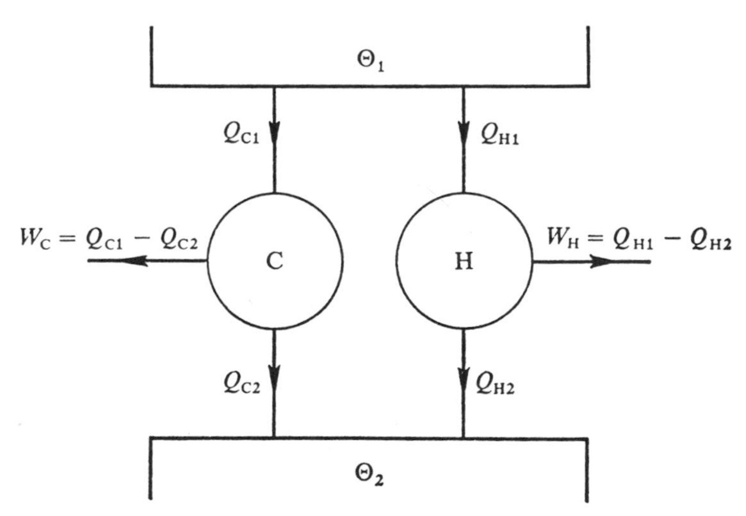

Fig. 4 . 5. The proof of Carnot's theorem.

That is,

$$W_C = W_H. \tag{4.3}$$

Then from (4.2)

$$Q_{C1} > Q_{H1}. \tag{4.4}$$

We now see that the composite engine consisting of C and H taken

[3] If we did not wish to adjust the *size* of the Carnot cycle, we could arrange for there to be no surplus mechanical energy by adjusting the relative *rates* of working of the engines as we did in the proof of the equivalence of the two forms of the second law. The argument would then follow through in exactly the same way.

together does no work but extracts heat from the cold reservoir and delivers an equal amount to the hot reservoir given by

$$Q_{C1}-Q_{H1}>0 \qquad \text{by (4.4).}$$

The composite engine therefore violates the Clausius statement of the second law, so that our hypothetical engine cannot exist and the theorem is proved.

Corollary. In Carnot's theorem we have shown that $\eta_{\mathrm{Carnot}} \geq \eta_{\mathrm{other}}$. If the hypothetical engine in the above proof were replaced by any reversible engine R, we should have shown that $\eta_C \geq \eta_R$. But now, since both engines are reversible, the Carnot engine could have been used to drive the other engine backwards giving the result $\eta_C \leq \eta_R$. These conditions may only both be satisfied if $\eta_C = \eta_R$. Thus we have shown that: *All reversible engines operating between the same reservoirs are equally efficient.*

It follows that the efficiency of any reversible engine operating between two reservoirs must be a function of the temperatures of the reservoirs only. That is, for *any* reversible engine,

▶
$$\frac{Q_1}{Q_2}=f\,(\Theta_1, \Theta_2) \qquad (4.5)$$

where f is a *universal* function of Θ_1 and Θ_2.

As mentioned earlier, the only reversible cycle in which a single working substance exchanges heat at two temperatures only is necessarily a Carnot cycle; but the second engine in the above argument may be as complex as we please (it might contain several subsidiary cyclic processes, for example) provided that it is reversible and exchanges heat with its surroundings via the two reservoirs only. A Carnot engine is the simplest engine fulfilling these conditions. An example of a more complicated one occurs in section 4.6.

4.6. Thermodynamic Temperature

From (4.5) we may derive thermodynamic temperature. Consider two Carnot engines, the first, C_1, operating between reservoirs at Θ_1 and Θ_2 and the second, C_2, operating between Θ_2 and Θ_3. Let C_1 absorb Q_1 at Θ_1 and reject Q_2 at Θ_2. Adjust the relative sizes of the cycles so that C_2 absorbs Q_2 at Θ_2 and rejects Q_3 at Θ_3 (Fig. 4.6). Then from (4.5) we have for C_1

$$\frac{Q_1}{Q_2}=f\,(\Theta_1, \Theta_2), \qquad (4.6)$$

and for C_2

$$\frac{Q_2}{Q_3} = f'(\Theta_2, \Theta_3). \tag{4.7}$$

But since no net heat is exchanged at Θ_2, the Θ_2 reservoir is superfluous. (The two engines could exchange the heat directly while performing exactly the same cycles.) Thus the Θ_2 reservoir may be bypassed while equations (4.6) and (4.7) remain unaffected. However, we may then consider the two Carnot engines to form a composite (reversible) engine exchanging heat at Θ_1 and Θ_3 only.[4] Applying (4.5) to the composite engine

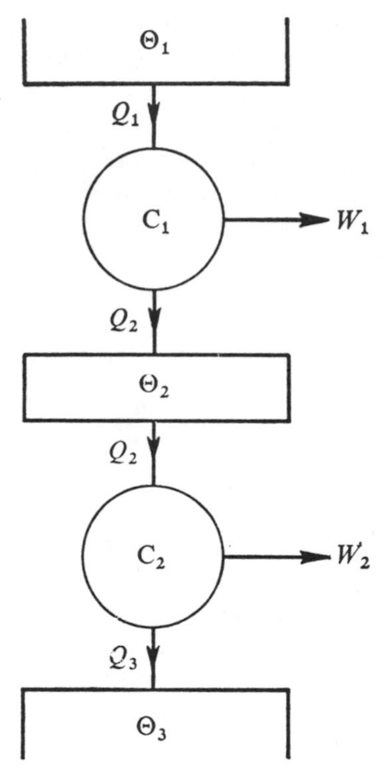

Fig. 4.6 The derivation of thermodynamic temperature.

$$\frac{Q_1}{Q_3} = f''(\Theta_1, \Theta_3). \tag{4.8}$$

[4] Alternatively, since the Θ_2 reservoir is unchanged by the operation of the engines, we could include it with them as part of the composite engine.

From (4.6), (4.7), and (4.8)

$$f''(\Theta_1, \Theta_3) = f(\Theta_1, \Theta_2) . f'(\Theta_2, \Theta_3).$$

But the left-hand side is independent of Θ_2, so that Θ_2 must cancel out from the terms on the right. This can only happen if the f's are of the form $T(\Theta_1)/T(\Theta_2)$ where the T's are universal quantities depending on the empirical temperatures only. Returning to equation (4.6), we may therefore put

$$\frac{Q_1}{Q_2} = \frac{T_1}{T_2} \tag{4.9}$$

which defines *thermodynamic temperature* apart from the constant of proportionality which fixes the size of the unit. In other words, thermodynamic temperature is defined so that *the ratio of the thermodynamic temperatures of two reservoirs is equal to the ratio of the amounts of heat exchanged at those reservoirs by a reversible engine operating between them.*

In section 8.2 we shall show that it is thermodynamic temperature which appears in the equation of state of the perfect gas. This is why measurements of thermodynamic temperature are usually ultimately based on gas thermometry. (See sections 2.4–6.)

In terms of thermodynamic temperature, the efficiency of a reversible engine operating between two reservoirs at temperatures T_1 and T_2 becomes

$$\eta = \frac{W}{Q_1} = 1 - \frac{Q_2}{Q_1} = 1 - \frac{T_2}{T_1}. \tag{4.10}$$

4.7. The Uniqueness of Reversible Adiabatics

In the various proofs we have given above, we have assumed that the path followed by a system in the course of a reversible adiabatic change is uniquely defined. In particular, we have assumed that if a system sets off along an adiabatic path from a particular state on one isotherm, it will always intersect another isotherm at the same point. If this were not the case, the Carnot cycle which the system performs would not be defined and the work done in the cycle would not be determined uniquely. The assumption that the adiabatics *are* uniquely defined amounts to assuming that there must be some function of state which is constant for (reversible) changes in which $đQ = 0$, the constancy of this quantity determining what states are accessible to the system. However, we know that

Q is not a function of state, and therefore the assumption we have made needs further justification.

One way out of this difficulty is simply to appeal to experimental evidence that reversible adiabatic processes *are* uniquely defined. However, it is easy to show that this must be so in the case of a two-parameter system.

The reversible adiabatic is the curve for which

$$đQ = \mathrm{d}U + p\,\mathrm{d}V = 0. \tag{4.11}$$

We choose p and V as independent variables and rearrange this equation into a linear differential form in $\mathrm{d}p$ and $\mathrm{d}V$. In terms of $\mathrm{d}p$ and $\mathrm{d}V$, $\mathrm{d}U$ becomes

$$\mathrm{d}U = \left(\frac{\partial U}{\partial p}\right)_V \mathrm{d}p + \left(\frac{\partial U}{\partial V}\right)_p \mathrm{d}V. \tag{4.12}$$

Substituting in (4.11)

$$\left(\frac{\partial U}{\partial p}\right)_V \mathrm{d}p + \left\{p + \left(\frac{\partial U}{\partial V}\right)_p\right\} \mathrm{d}V = 0 \tag{4.13}$$

which is the differential equation which the system parameters must obey for a reversible adiabatic change. Now the coefficients of $\mathrm{d}p$ and $\mathrm{d}V$ in (4.13) are functions of state so that the direction of an adiabatic change is uniquely determined by an equation of the form

$$F_1(p, V)\,\mathrm{d}p + F_2(p, V)\,\mathrm{d}V = 0. \tag{4.14}$$

Equation (4.13) may therefore be integrated to give unique adiabatics. The proof does not, however, generalize to cases where there are more than two independent variables.[5]

Thus, for the moment, we may only take adiabatics to be unique

[5] If one thinks of $\mathrm{d}p$ and $\mathrm{d}V$ as infinitesimal vectors in p–V space, then equation (4.14) simply states that any permitted infinitesimal change, represented by the vector $(\mathrm{d}p, \mathrm{d}V)$, must be perpendicular to the vector $(F_1(p, V), F_2(p, V))$ since their scalar product is zero. Since the coordinate space has two dimensions only, this condition defines a unique line. With three degrees of freedom, it defines a plane containing the initial state, and the condition is satisfied by any line lying in that plane. With this extra degree of freedom it would, in general, be possible to pass from any point in system coordinate space to any other while always satisfying the adiabatic condition at all points of the path. All states of the system would then be mutually accessible by adiabatic paths and there would be no unique adiabatic surfaces. We shall postpone demonstration of the general result until we have introduced entropy in chapter 5.

for systems with two degrees of freedom, and the proofs we have given earlier using Carnot engines are only valid in this case. However, when we come to discuss general cyclic processes, we shall be able to derive results from them which are valid for many-parameter systems. Eventually we shall show that there *is* a function of state, entropy, which is conserved in reversible adiabatic processes. It then follows immediately that adiabatic surfaces exist for many-parameter systems, since they are surfaces on which the function of state, entropy, is constant. (See sections 5.2 and 6.2.)

Incidentally, we have not shown that an adiabatic may not cut an isothermal more than once; but this is immediately precluded by the second law, for an engine based on a system in which this occurs would violate the Kelvin statement.

4.8. Refrigerators and Heat Pumps

In the proof of Carnot's theorem, we used the reversible property of a Carnot engine. By the expenditure of mechanical work it was made to run backwards, extracting heat from a cold reservoir and rejecting it to a hot one (Fig. 4.7). Any device which, by the use of mechanical work, transfers heat continually from colder to hotter is called a *refrigerator* or a *heat pump*.

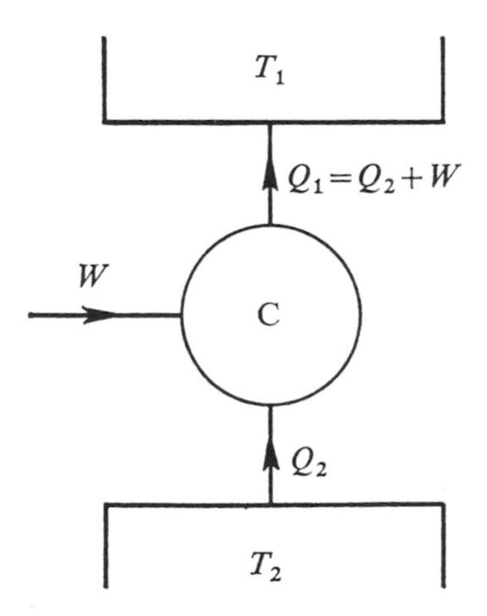

Fig. 4.7. A Carnot engine driven backwards absorbs heat at the cold reservoir and rejects heat at the hot.

Refrigerators. The function of a refrigerator is to extract heat from a body which is at a lower temperature than the surroundings. The efficiency of a refrigerator should therefore be defined in terms of the amount of heat extracted for a given expenditure of mechanical work. For a perfect refrigerator, using a Carnot engine,

$$\eta_r = \frac{Q_2}{W} = \frac{T_2}{T_1 - T_2}. \tag{4.15}$$

The efficiency of the ideal refrigerator is shown in Fig. 4.8. For moderate degrees of cooling the efficiency is high. Down to $T_2/T_1 = 0{\cdot}5$ more heat is absorbed than work required; but for a given extraction of heat the work required becomes very large as the temperature ratio is increased. For a domestic refrigerator the upper temperature

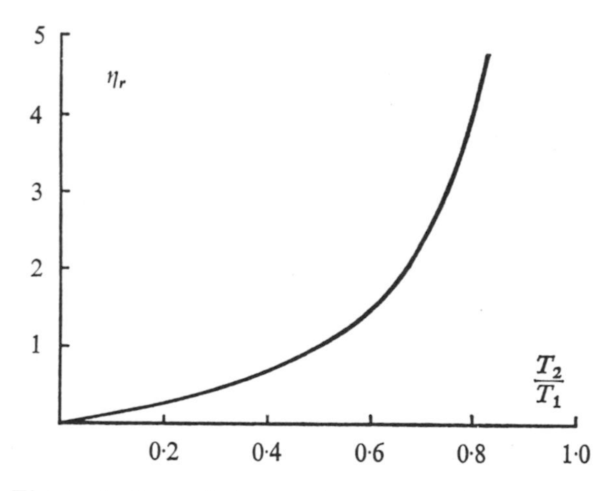

Fig. 4.8. The efficiency of an ideal refrigerator.

is usually kept close to room temperature by exchange of heat with the surroundings through cooling fins, while the lower temperature is kept somewhat below freezing point. With $T_1 = 300$ K and $T_2 = 270$ K, $\eta = 9$. On the other hand, to absorb 4 watts of heat at 1 K with a refrigerator working from room temperature would require more than 1 kW of power. (This shows why it becomes increasingly difficult to obtain cooling at very low temperatures.) In practice, efficiencies will fall well below these ideal figures.

Heat Pumps. The function of a heat pump is to deliver heat to some body which is at a higher temperature than its surroundings. The efficiency of a heat pump should therefore be defined in terms of the amount of heat delivered at the higher temperature for a given expenditure of mechanical work. For a perfect heat pump, using a Carnot engine,

$$\eta_p = \frac{Q_1}{W} = \frac{T_1}{T_1 - T_2} = 1 + \eta_r. \qquad (4.16)$$

Figure 4.9 shows the efficiency of an ideal heat pump. For small temperature differences considerably more heat is supplied than power consumed. This makes the heat pump very attractive as a device for heating buildings, heat being extracted from the surrounding atmosphere and delivered to the building at a little above room temperature. Taking $T_1 = 320$ K and $T_2 = 280$ K, 8 kW

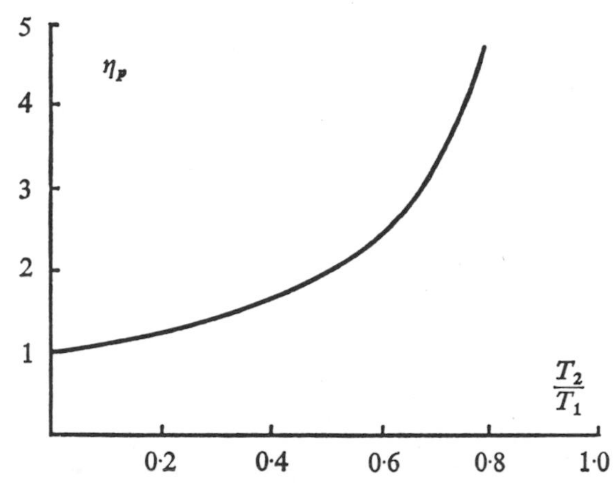

Fig. 4.9. The efficiency of an ideal heat pump.

of heating would involve a power consumption of only 1 kW. Unfortunately, the high cost and low efficiency of any practical plant make this method of heating of doubtful economic advantage and it has rarely been used. As the temperature ratio increases the potential advantages of the heat pump becomes smaller. When $T_2 \to 0$ the heat delivered becomes equal to the work required and the heat pump has no advantage over a device which turns the work directly into heat such as a simple electric heater.

4.9. Real Heat Engines

In completely general terms, little can be said about the efficiencies of real heat engines although a crude comparison with a Carnot cycle is sometimes possible. If we know the extremes of temperature involved in the cycle of the real engine, then certainly its efficiency must be less than that of a Carnot engine operating between reservoirs at these extremes. Such a simple comparison is sufficient to show why the early steam engines were so inefficient. Steam was available at somewhat above atmospheric pressure, say at 390 K, and was condensed by water at a temperature somewhat below normal boiling point, say 350 K. The efficiency of a Carnot engine operating between these temperatures would only be 10 per cent and, of course, for the steam engines it was much smaller still. In the modern steam engine, the efficiency has been improved by using high pressure steam and forcing T_1 up; but the steam engine is still an inefficient means of generating mechanical power from heat because of the comparatively limited temperature range which is practicable. In contrast, one would expect the internal combustion engine to be capable of much higher efficiencies because of the extremely high temperatures which are involved in the explosion.

To discuss a real heat engine in any detail it is always necessary to invent an idealized cycle which may be used as a reasonable representation of the cycle of the real engine. Calculations based on such an idealized cycle will give an upper limit to the efficiency of the real heat engine. The idealization involves two basic approximations. The first is that the working substance is a single pure substance. In the case of the internal combustion engine, this is clearly far from the truth. The working substance is, in fact, a mixture of gases and vapours and its composition changes during the cycle. For the internal combustion engine, air is usually chosen to represent the working substance. Cycles based on air are known as 'air standard cycles'.

The second approximation consists of replacing the real cycle by a reversible cycle. Again, this is clearly far from the truth. Most real cycles proceed rapidly and the conditions are far from quasi-static: heat flows through finite temperature gradients; there is friction and turbulence. In practice, it is usually this approximation which introduces the greatest errors.

We shall illustrate the use of an idealized cycle by considering one example, that of the petrol engine.

4.9.1. The Petrol Engine

In the petrol engine, the cycle consists of six parts. Four of these involve motion of the piston and are called *strokes*. The cycle proceeds as follows:

1. *Intake stroke.* The mixture of petrol and air is drawn into the cylinder through the intake valve by the movement of the piston.

2. *Compression stroke.* The intake valve now closes and the piston moves up the cylinder compressing the mixture rapidly. The compression is nearly adiabatic and there is a considerable temperature rise.

3. *Explosion.* When fully compressed, the mixture is caused to explode. There is negligible movement of the piston during the explosion so that the volume remains unchanged, but a very high temperature and pressure are reached.

4. *Power stroke.* The hot combustion products expand, doing mechanical work on the piston. There is a considerable drop in pressure and temperature.

5. *Valve exhaust.* At the end of the power stroke the exhaust valve opens. The combustion products which are still at a high pressure flow out rapidly into the atmosphere. There is a sudden drop in pressure.

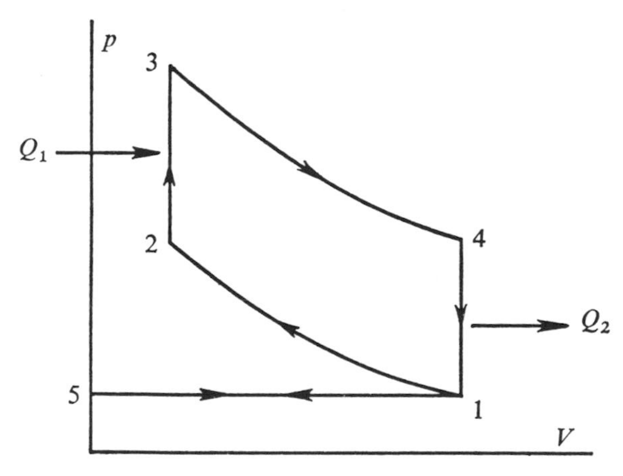

Fig. 4.10. The air standard Otto cycle.

6. *Exhaust stroke.* The piston moves up the cylinder forcing the remaining gases out into the atmosphere. The exhaust valve then closes and the intake valve opens in readiness for the next intake stroke.

The petrol engine cycle is clearly highly irreversible. The idealized cycle which replaces it is known as the Air Standard Otto Cycle and is illustrated in Fig. 4.10. Air is taken as the working substance and is assumed to obey the ideal gas laws (see section 8.2) with constant specific heats. All processes are assumed to be reversible. The various parts of the petrol engine cycle are then represented as follows:

$5 \to 1$. *The intake stroke.* A quasistatic isobaric intake of air at p_0 to a volume V_1.

$1 \to 2$. *The compression stroke.* A quasistatic adiabatic compression from V_1 to V_2 during which the temperature rises from T_1 to T_2 according to the perfect gas equation,

$$T_1 V_1^{\gamma-1} = T_2 V_2^{\gamma-1} \tag{4.17}$$

where γ is the ratio of the specific heats.

$2 \to 3$. *The explosion.* A quasistatic isovolumic rise of temperature and pressure brought about by the absorption of heat from a series of reservoirs between T_2 and T_3.[6]

$3 \to 4$. *The power stroke.* A quasistatic adiabatic expansion producing a drop in temperature according to

$$T_3 V_2^{\gamma-1} = T_4 V_1^{\gamma-1}. \tag{4.18}$$

$4 \to 1$. *Valve exhaust.* A quasistatic isovolumic drop of temperature to T_1 (and of pressure to p_0) brought about by exchange of heat with a series of reservoirs between T_4 and T_1.

$1 \to 5$. *Exhaust stroke.* A quasistatic isobaric expulsion of the air.

Clearly, the two isobaric processes $5 \to 1$ and $1 \to 5$ cancel one another out, and in calculating the efficiency we only need to consider the rest of the cycle.

The heat absorbed along $2 \to 3$ is $Q_1 = \int_{T_2}^{T_3} C_v \, dT = C_v(T_3 - T_2)$,

and rejected along $4 \to 1$ is $Q_2 = \int_{T_4}^{T_1} C_v \, dT = C_v(T_4 - T_1)$.

Applying the first law, the efficiency is

[6] The series of reservoirs is necessary so that no temperature difference ever occurs between the system and the reservoir supplying heat. If a temperature difference were set up, the flow of heat would then become thermodynamically irreversible.

$$\eta = \frac{W}{Q_1} = \frac{Q_1 - Q_2}{Q_1} = 1 - \frac{(T_4 - T_1)}{(T_3 - T_2)}.$$

From (4.17) and (4.18)

$$(T_4 - T_1)\, V_1^{\gamma-1} = (T_3 - T_2)\, V_2^{\gamma-1}.$$

Hence

$$\eta = 1 - \left(\frac{V_2}{V_1}\right)^{\gamma-1} = 1 - \frac{1}{r^{\gamma-1}} \qquad (4.19)$$

where r is called the *compression* (or expansion) *ratio.*

To obtain the highest efficiency, the expansion ratio has to be as large as possible. However, it cannot be made too large because eventually regions in the fuel mixture detonate during the combustion rather than burning smoothly. The resulting *pinking* or *knocking* is mechanically bad for the engine and also reduces the efficiency. With modern fuels a compression ratio of about 10·5 can be used. Taking $\gamma = 1{\cdot}4$ for air, this gives a theoretical maximum efficiency of 61 per cent. In a real engine, the efficiency probably only reaches half this value.

5. Entropy

5.1. Clausius' Theorem

So far, we have only discussed cycles in which the system exchanges heat at two temperatures only. For heat engines based on such cycles we have, according to Carnot's theorem,

$$\eta \leq \eta_{\mathrm{rev}}$$

where η_{rev} is the efficiency of a reversible engine operating between the same temperatures. Substituting for the efficiencies,

$$1-\frac{Q_2}{Q_1} \leq 1-\frac{Q_{r2}}{Q_{r1}}$$

$$\frac{Q_2}{Q_1} \geq \frac{Q_{r2}}{Q_{r1}} = \frac{T_2}{T_1}$$

by the definition of absolute temperature.
Therefore,

$$\frac{Q_2}{T_2} \geq \frac{Q_1}{T_1}.$$

Taking the heat entering the system as positive, we may write this

$$\Sigma \frac{Q}{T} \leq 0.^{[1]} \qquad (5.1)$$

We shall now prove a corresponding result for general cyclic processes of any degree of complexity. In particular, there will be no restrictions on the number of degrees of freedom of the system nor on the temperature at which it may exchange heat with its surroundings.

To make the system execute the cycle, an appropriate series of adjustments has to be made to its parameters (involving work) and at each stage the appropriate amount of heat has to be supplied. The cycle itself may be irreversible, but we may supply the heat reversibly by operating a minute Carnot engine between the system

[1] It should be noted that the inequality becomes stronger as the engine becomes less efficient and less reversible. We shall later see how this comes about in a more general way.

and a large reservoir at constant temperature (Fig. 5.1). By ensuring that the Carnot engine is in thermal equilibrium with the reservoir or system during the transfer of heat, no irreversibility is involved there. The series of processes followed by the Carnot engine C is as follows. (All are reversible.)

(a) C is at T_0.
(b) C is compressed (or expanded) adiabatically until its temperature is T.
(c) C is placed in contact with the system and absorbs or supplies heat by an isothermal change at T.
(d) C is expanded (or compressed) adiabatically until its temperature is T_0.
(e) C is placed in contact with the reservoir and compressed (or expanded) isothermally at T_0 until it regains its original state.

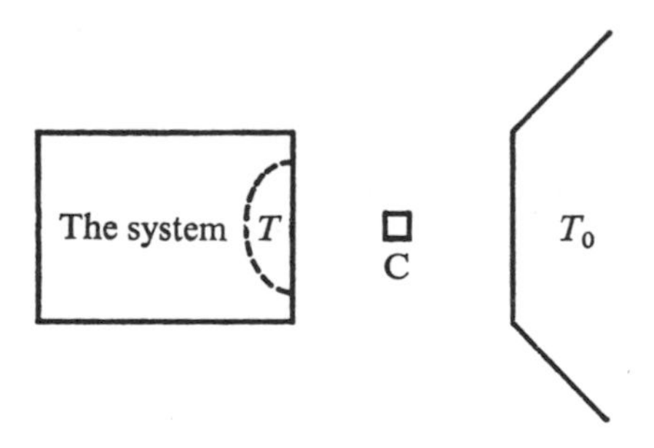

Fig. 5 . 1. The proof of Clausius' theorem.

In this way, the complex engine executes its cycle in infinitesimal steps and no assumptions are made about the uniqueness of its adiabatics nor about whether the working substance can depart from the specified cycle.[2]

If the heat supplied to the working substance at T in one journey of the Carnot engine is $đQ$, the corresponding heat absorbed from

[2] A misleading method which is often used to prove Clausius' theorem is to superimpose on the general cycle a mesh of adiabatics and isothermals so as to subdivide it into infinitesimal cycles to which (5.1) is applied. There are two objections to this. In the first place, the argument depends on the assumption that the system could exist in all of the states involved in the subdivision. Clearly, this might not be true. The second, and more serious objection is that we have not yet proved the existence of unique adiabatic surfaces for systems of more than two variables. Thus the result would not be of general validity.

he reservoir is

$$\frac{T_0}{T}đQ.$$

Ience, the heat absorbed from the reservoir in one complete cycle f the complex engine is

$$T_0 \oint \frac{đQ}{T} \leq 0,$$

y the Kelvin statement of the second law. However, T_0 is necesarily positive, and therefore,

$$\oint \frac{đQ}{T} \leq 0 \qquad \textit{for any cycle.} \tag{5.2}$$

If the complex cycle were *reversible* we could have executed it in he opposite direction and derived the result

$$\oint \frac{đQ}{T} \geq 0. \tag{5.3}$$

But (5.2) applies to *any* cycle and therefore necessarily to reversible cycles in particular. Hence, if the cycle is reversible both (5.2) and (5.3) must be satisfied, giving

$$\oint \frac{đQ}{T} = 0 \qquad \textit{for a reversible cycle.} \tag{5.4}$$

The two results (5.2) and (5.4) together form *Clausius' theorem*, which may be stated formally as follows:

For any closed cycle, $\oint \frac{đQ}{T} \leq 0$, *where the equality necessarily holds for a reversible cycle.*

Clausius' inequality is very important, for our whole treatment of irreversible processes will follow from it.

It is important to be clear about the significance of T in the above results. In an irreversible cycle the various parts of the system might not always be in equilibrium with one another and, in particular, there might be temperature differences, making it impossible to define a temperature for the system as a whole. In the proof, T is the temperature of the Carnot engine as the heat is transferred across the boundary of the system. Thus the T appearing in the integrals is the *temperature at which heat is supplied to the system.* Only if the source of heat is in thermal equilibrium with the system as a whole does it become the temperature of the system also.

5.2. Entropy

We now define a new variable, the entropy S, by the relation

$$dS = \frac{đQ}{T}$$

for an infinitesimal **reversible** *change.* To emphasize that the equality holds for reversible changes only, the definition of S is written

$$dS = \frac{đQ_{rev}}{T}. \tag{5.5}$$

Then for a finite reversible change of state, the change in entropy is given by

$$S_2 - S_1 = \int_1^2 \frac{đQ_{rev}}{T}. \tag{5.6}$$

We shall now show that entropy is a function of state.

Proof that S is a Function of State

Construct any reversible cycle and select any two states A and B on it (Fig. 5.2). Then Clausius' theorem states

$$0 = \oint_{ACBDA} \frac{đQ_{rev}}{T} = \oint_{ACBDA} dS$$

from the definition of S, all processes being reversible.

$$= \int_{ACB} dS + \int_{BDA} dS.$$

Therefore,

$$\int_{ACB} dS = \int_{ADB} dS = S_B - S_A.$$

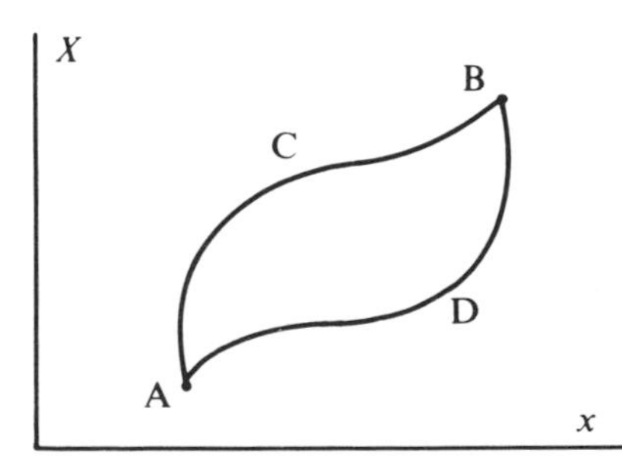

Fig. 5.2. The proof that entropy is a function of state.

If the path via D is kept fixed and the path via C varied, we see that

$$\int_{\mathrm{ACB}} \mathrm{d}S = S_{\mathrm{B}} - S_{\mathrm{A}}$$

always takes the same value for *any reversible path* from A to B. Hence, apart from an arbitrary additive constant, S must be uniquely defined for every state of the system: i.e., *S is a function of state.*

Since S is a function of state, $\mathrm{d}S$ must be a perfect differential (i.e., it is uniquely defined for any given change of state and can therefore always be integrated). But we have defined S by the equation

$$\mathrm{d}S = \frac{đQ_{\mathrm{rev}}}{T}$$

where $đQ$ is *not* a perfect differential. Thus we have discovered that there is an *integrating factor* for $đQ_{\mathrm{rev}}$, namely, $1/T$.

It also follows immediately that adiabatics exist and are unique for systems of any number of degrees of freedom, for they are simply the surfaces of constant entropy, the *isentropes*. This deals with the point we had to leave in section 4.7.

5.3. Entropy in Irreversible Changes

Since entropy is a function of state, the change in entropy accompanying a given change of state must always be the same *however the change in state occurs*. Only when the change takes place reversibly, however, is the entropy change related to the heat transfer by the equation

$$\Delta S = \int \frac{đQ}{T},$$

for we imposed the condition of reversibility in the initial definition. What is the relationship between entropy change and heat transfer in irreversible processes?

Consider an irreversible change, A→B. Construct any reversible path R between A and B, thus forming an irreversible cycle ABRA (Fig. 5.3). For the irreversible cycle Clausius' theorem gives

$$\oint \frac{đQ}{T} \leq 0.$$

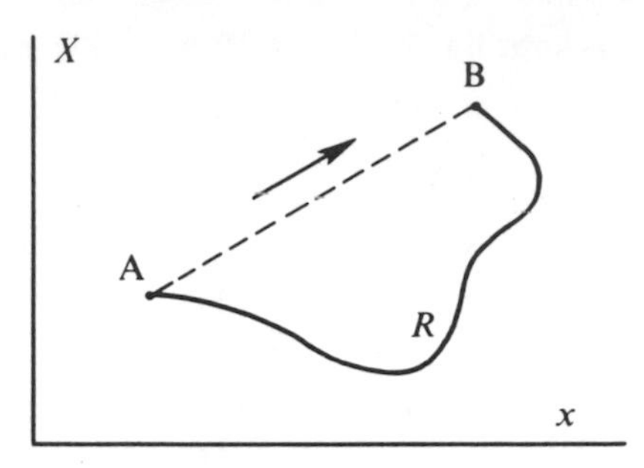

Fig. 5.3. The determination of the behaviour of entropy in an irreversible change.

Taking the integral in two parts,

$$\int_{A_{\text{irrev.}}}^{B} \frac{đQ}{T} + \int_{B_{\text{rev.}}}^{A} \frac{đQ}{T} \leq 0$$

i.e.,

$$\int_{A_{\text{irrev.}}}^{B} \frac{đQ}{T} \leq \int_{A_{\text{rev.}}}^{B} \frac{đQ}{T}.$$

But

$$\int_{A_{\text{rev.}}}^{B} \frac{đQ}{T} = S_B - S_A$$

by definition of entropy.
Thus

$$\int_{A_{\text{irrev.}}}^{B} \frac{đQ}{T} \leq S_B - S_A,$$

or,

$$dS \geq \frac{đQ}{T}$$

for a differential irreversible change.
Thus, we have the general result

$$\blacktriangleright \qquad dS \geq \frac{đQ}{T} \tag{5.7}$$

for *any* infinitesimal change where *the equality necessarily applies if the change is reversible*. Again, T is the temperature at which the heat is supplied to the system. Only when the source of heat is in thermal equilibrium with the system as a whole does it become the temperature of the system also.

Equation (5.7) is extremely important. It contains all the information required for dealing with efficiency and irreversibility in thermal processes. It may therefore be thought of as the focal point of the second law since it is through it that the objectives of the second law are realized.

For a system which is thermally isolated (or completely isolated) $đQ=0$. Applying (5.7) we see that $dS \geq 0$. This general result is known as the *Law of Increase of Entropy* which may be stated formally as: *The entropy of an isolated system cannot decrease.* A particular application of this law is that it may be used to determine the equilibrium configuration of an isolated system, for in approaching equilibrium its entropy can only increase. The final equilibrium configuration is thus that for which the entropy is as large as possible. Later, when we come to discuss the interpretation of entropy, we shall see how this principle may be applied.

It should be noted that the law of increase of entropy provides a natural direction to the time sequence of natural events. Within the mechanistic framework of Newtonian mechanics all processes are reversible in time. (The equations remain unaltered in form on replacing t by $-t$.) Why then is there the inevitable sequence to events, the so called 'arrow of time'? Thermodynamics does not answer this problem but it does provide a new insight. The natural direction of events is that in which entropy increases. All changes are part of the irreversible progress towards universal equilibrium. Thus, the arrow of time results from there *not* being thermodynamic equilibrium throughout the universe. As long as temperature differences or density differences exist natural evolution will continue and events will be directed forwards towards equilibrium.

5.4. The Entropy Form of the First Law

From the first law we were able to deduce the existence of internal energy U, a function of state. For any change of state, *however it occurs*, the change in U is given by equation (3.3), namely:

$$dU = đQ + đW$$

where $đQ$ and $đW$ are not differentials of functions of state and are therefore not individually defined for a given change of state.

To separate the contributions to U from heat and work the constraints on the system have to be known so that the path of the change may be found. If the change takes place reversibly, the work done may be expressed in terms of the system's parameters of state in the form $\sum X_i \, dx_i$, and only when the path is known can this be integrated. Thus, taking a simple fluid as our model, we have

$$dU = đQ + đW \qquad \textit{always} \tag{5.8}$$

$$đW = -p \, dV \qquad \textit{for reversible changes} \tag{5.9}$$

and we have defined entropy such that

$$đQ = T \, dS \qquad \textit{for reversible changes.} \tag{5.10}$$

Substituting (5.9) and (5.10) in (5.8) we obtain

$$dU = T \, dS - p \, dV \quad \textit{for reversible changes.} \tag{5.11}$$

However, in this equation *all the variables are functions of state* so that all the differentials are perfect. As a result, integration of this equation must be independent of the path of integration and the equation may be applied to any change of state, *however accomplished*. To use the equation, we only require that the initial and final states be defined. We may then choose any path for integration between them which is convenient for calculation. In particular, by choosing some reversible path we may now calculate the changes in internal energy accompanying irreversible changes. Thus we have succeeded in expressing dU in terms of state functions only, and we have

▶ $$dU = T \, dS - p \, dV \qquad \textit{always.} \tag{5.12}$$

For irreversible changes, the equalities (5.9) and (5.10) do not hold. We have already shown that in this case (5.10) becomes the inequality $đQ \leq T \, dS$ so that for (5.12) to remain true $đW \geq -p \, dV$. This is what one would expect. In the presence of irreversibility (when there is friction, for example), the total work done is greater than that which would be required to effect the same change in volume of the system without the irreversibility.

The general form of the first law is thus

$$dU = T \, dS + \sum X_i \, dx_i \tag{5.13}$$

where X_i and x_i are the intensive variables and their conjugate extensive variables. It is clear from its definition, that entropy is an *extensive* variable so that from the form of (5.13) thermodynamic temperature must be its corresponding intensive variable. The term $T\,dS$ is thus entirely similar to the work terms and may be grouped with them. This gives the first law in a condensed form:

$$dU = \sum X_i \, dx_i \tag{5.14}$$

where the summation necessarily includes the term $T\,dS$ which is relevant to all systems.

5.5. Entropy and the Degradation of Energy

The work that can be extracted *from* a system in an infinitesimal change of state is $đW = đQ - dU$. We have shown that $đQ$ is related to the entropy change by $đQ \leq T_0\,dS$, where T_0 is the temperature at which the heat is supplied, so that $đW$ must satisfy the inequality $đW \leq T_0\,dS - dU$. Thus, for a given change of state (so that dU and dS are fixed), the maximum amount of work is extracted from the system when the equality applies; that is, when the change is reversible. In this case, the total entropy change of the system and its surroundings is zero,[3] for, in any process involving reversible exchange of heat with the surroundings, $dS_{\text{system}} = -dS_{\text{surroundings}}$; whereas, in an irreversible change, the entropy change of the surroundings (assuming no irreversibility there) is $dS_0 = -đQ/T_0$, while that of the system satisfies the inequality $dS \geq đQ/T_0$. In this case, the entropy of the universe may increase, and if it does, we are able to extract less work from the system than would have been the case if the same change had been made reversibly, Thus, associated with the increase of entropy is the 'loss' of some energy which could have been used for work. Clearly, this energy does not vanish, for this would violate the first law, but rather it takes a form from which it may be converted into work with less efficiency than previously. The energy becomes *degraded* in that it is less useful for work. We may illustrate this by a simple example.

Consider two bodies 1 and 2 which are at temperatures T_1 and T_2. Suppose that $T_1 > T_2$. Then if we connect the bodies together by a thermal resistance and allow a small quantity of heat q to flow, the total change of entropy is

$$\Delta S = \Delta S_1 + \Delta S_2 = q\left(\frac{1}{T_2} - \frac{1}{T_1}\right) > 0, \text{ while } T_1 > T_2.$$

[3] In this kind of situation, one often speaks of the entropy of the universe as being conserved. This is simply a convenient way of lumping together the system and its surroundings.

Thus, the entropy increases and will continue to increase as long as the heat flows, bringing the bodies towards equilibrium.

Now suppose that instead of allowing q to flow from 1 to 2 we used it to operate a Carnot engine from which to obtain mechanical work. Let us suppose that T_0 is the temperature of the coldest reservoir we have to hand for use with the Carnot engine. Then by extracting q from 1 we could have obtained work

$$W_1 = q\left(1 - \frac{T_0}{T_1}\right).$$

If, however, we first allow q to flow from 1 to 2 and then use it to operate the Carnot engine we only obtain

$$W_2 = q\left(1 - \frac{T_0}{T_2}\right) < W_1.$$

Thus, in the course of the irreversible heat conduction the energy has become degraded to the extent that the useful work we may obtain from it has been decreased by

$$\Delta W = W_1 - W_2 = T_0 \Delta S.$$

The increase in entropy in an irreversible change is thus the measure of the extent to which energy becomes degraded in that change. Conversely, in order to extract the maximum amount of useful work from a system or set of systems, the changes must be performed in a reversible manner so that the total entropy (entropy of the system and its surroundings) is conserved.

It is worth pointing out that if the two bodies in the above illustration were allowed to reach thermal equilibrium (a) by heat conduction and (b) by operating a Carnot engine between them and extracting work, the final equilibrium temperatures would be different in the two cases. In the first, $U_1 + U_2$ is conserved and the final temperature is

$$T_f^{(U)} = (C_1 T_1 + C_2 T_2)/(C_1 + C_2)$$

where the C's are the thermal capacities, which, for simplicity, we have taken to be constants. In the second case, $S_1 + S_2$ is conserved and $W = -\Delta(U_1 + U_2)$. In the isentropic process, the final temperature is given by

$$T_f^{(S)} = T_1^{[C_1/(C_1+C_2)]} T_2^{[C_2/(C_1+C_2)]} < T_f^{(U)}. \qquad (5.15)$$

The difference in the final temperature corresponds to the lower value for the total internal energy which results from work having been done.

5.6. Entropy and Order

We have shown that the equilibrium state of an isolated system is that for which the entropy takes on its maximum value, so that in terms of macroscopic variables the maximization of entropy is the condition for determining the equilibrium configuration. An alternative approach would be to apply probability theory at the microscopic level to the various possible configurations of the system and to seek that configuration whose probability is greatest. This is the method of the discipline known as *Statistical Mechanics* or *Statistical Thermodynamics*.[4] The exact definition of the statistical probability of a particular macroscopic state, for which we shall use the symbol g, is outside the scope of this book, but its relationship to entropy is so important in making it possible to link macroscopic and microscopic properties that some discussion of it is essential.

In seeking the most probable configuration of a system we are, in fact, seeking the configuration of the greatest *disorder* permitted by the constraints to which the system is subjected. A configuration which requires particular conditions of order (such as that no molecules should be in a particular region of space), is clearly less likely to occur spontaneously than one in which no conditions are specified. Thus the most probable configuration, the equilibrium configuration, is that in which the disorder is as great as possible. The statistical probability of a particular configuration is therefore a measure of its disorder. Without involving ourselves in the exact definition of g we may illustrate its connection with disorder by taking a simple example.

Consider a fixed mass of gas in a container. We divide the container into two equal parts, 1 and 2, and consider the probability that the molecules will all be in one half. The probability that a particular molecule will be in 1 is clearly $\frac{1}{2}$. The probability of finding two particular molecules in 1 at the same time is $\frac{1}{2}\times\frac{1}{2}$. Extending the argument to all N molecules, the probability that all molecules will be in 1 at any particular time is $(\frac{1}{2})^N$. We may therefore compare the statistical probability that all the molecules are in 1, g_1, with that for the molecules to occur randomly throughout the whole box, g_{1+2}:

$$\frac{g_1}{g_{1+2}} = (\tfrac{1}{2})^N. \tag{5.16}$$

[4] See F. Mandl, *Statistical Physics*, Wiley, 1971; or C. Kittel, *Thermal Physics*, Wiley, 1969.

(If the box contains 1 mol of the gas, we have $N=6\times10^{23}$ and we see that the chance of finding all the gas in one half of the box is about 1 in $10^{1.8\times10^{23}}$. This would occur spontaneously about once in $10^{1.8\times10^{23}}$ universes: a rare event.)

This simple illustration demonstrates the connection between statistical weight and disorder. We have shown that for equilibrium the macroscopic quantity entropy must be maximized and how the corresponding microscopic condition is the maximization of g, which is related to the disorder in the system. Can we arrive at an explicit connection between entropy and order? We may see what this might be by considering two systems, 1 and 2. Entropy is an extensive variable (section 5.4), so that the total entropy of the two systems taken together is

$$S_{1+2}=S_1+S_2. \tag{5.17}$$

The probabilities of finding each system in the particular configuration we specify for it at the same time is the product of the probabilities for each system alone:

$$g_{1+2}=g_1\,g_2. \tag{5.18}$$

Clearly, (5.17) and (5.18) are satisfied simultaneously if

$$S=k\ln g,$$

where k is a constant.
We may prove that this is necessarily the form of the relation as follows:
Suppose

$$S=f(g).$$

Then according to (5.17) and (5.18)

$$f(g_1g_2)=f(g_1)+f(g_2).$$

Differentiating twice, with respect to first g_1 and then g_2,

$$f'(g_1g_2)+g_1g_2f''\,(g_1g_2)=0$$

or,

$$gf''(g)=-f'(g).$$

Integrating,

$$\ln f'(g)=-\ln\,(g)+\text{const.}$$

or

$$f'(g)=k/g$$

where k is a constant.

Therefore,

$$f(g) = k \ln g + g_0$$

or

$$S = k \ln g + S_0$$

where S_0 is the constant of integration which it is convenient to take as zero corresponding to a statistical probability of unity for a completely ordered state. Thus we have proved that the relation between the entropy and the statistical probability is

$$\blacktriangleright \qquad S = k \ln g. \qquad (5.19)$$

This is the important *Boltzmann relation* which links classical thermodynamics with the microscopic properties of the system. We may show that k is Boltzmann's constant, R/N_A, by considering again the perfect gas contained in a box. We calculate the difference in entropy between the state in which the gas is entirely in one half of the box, and that in which it is uniformly distributed throughout the box. We do this by first imagining that the gas is constrained to one half of the box by a partition and that the partition is then punctured to allow the gas to fill the whole box. In the (irreversible) expansion $đQ = đW = 0$.
Therefore,

$$dU = TdS - pdV = 0.$$

We may now choose a convenient reversible path by which to evaluate the terms in the latter equation since all these are functions of state.
For $dU = 0$,

$$dS = \frac{p}{T} dV.$$

Using the perfect gas law, equation (8.10),

$$\frac{p}{T} = \frac{R}{V},$$

giving

$$\Delta S = R \int \frac{dV}{V} = R \ln \frac{V_2}{V_1} = R \ln 2 = kN_A \ln 2$$

where k is Boltzmann's constant and N_A the Avogadro constant. Comparing with (5.16) and (5.19) we see that k in (5.19) is indeed Boltzmann's constant.

Thus the entropy of a system is a measure of the disorder within it. This now makes it possible to interpret the degradation of energy discussed in the previous section. If energy is to be extracted from a system as efficiently as possible, that energy should be stored in an ordered form. A mechanical storage device such as a spring is ideal but thermal energy is also useful, particularly if the temperature is high; for T is the intensive variable coupled with S. When energy is degraded in an irreversible change it takes a less ordered form. This is obvious in the case of mechanical friction where ordered mechanical energy is dissipated as the disordered molecular motions of heat; but it applies also to the flow of heat down a temperature gradient where the non-equilibrium ordering of thermal energy, corresponding to the existence of the temperature difference, is reduced.

The direct relationship of entropy to disorder is extremely important in providing a link between macroscopic variables and microscopic processes. We shall illustrate this relationship with a few simple examples.

5.6.1. Specific Heats

Thermal energy is stored in a solid in the thermal motions of its atoms, and, if it is a metal, of its electrons also. The equations governing the motions of the atoms and electrons vary little with temperature, but the extent of the thermal motions increases as the temperature rises. The greater the thermal motions, the greater will be the microscopic disorder in the system, and the greater the entropy, the change in entropy being brought about by the heat which flows into the body as the temperature is raised. Thus, the common specific heats are associated with the *gradual* increase in disorder which accompanies a rise in temperature.

Now the heat capacities may be written in terms of entropy derivatives. For example,

$$C_V = \frac{đQ_V}{dT} = T\left(\frac{\partial S}{\partial T}\right)_V.$$

We may use this to calculate how the entropy of a solid varies with temperature.

The specific heat of an insulating solid follows the Debye law, according to which, at low temperatures, $C_V \propto T^3$, and at high temperatures, $C_V = \text{const.}$ (in agreement with Dulong and Petit's law). Thus, in the low temperature limit, S increases as T^3, and, at high temperatures, S varies as $\ln T$. In the case of a metal, the electronic contribution to the heat capacity is proportional to temperature so that the electronic contribution to the entropy is also proportional to T.

5.6.2. Specific Heat Anomalies

We would normally expect the thermal motions of the atoms of a material to increase smoothly as the temperature rises. In some substances it is found that superimposed on the smoothly varying background specific heat is an extra contribution which occurs at a particular temperature in the form of a relatively narrow peak. Such behaviour is known as a *specific heat anomaly*. The rapid rise

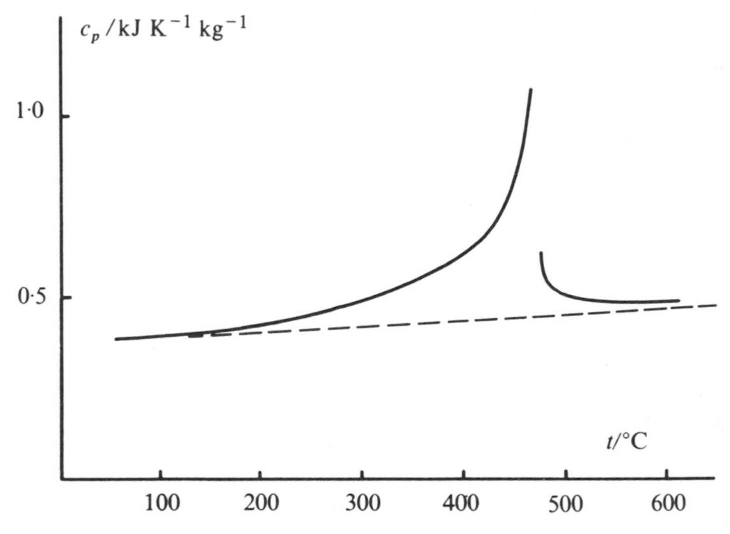

Fig. 5 . 4. The specific heat of β-brass.[5]

in entropy associated with the specific heat anomaly indicates that some microscopic change in order is occurring, and the magnitude of the entropy rise can be used as a guide to what the microscopic changes might be. At temperatures below the anomaly some aspect of the system must be ordered and above the anomaly disordered. Such a change is therefore known as an *order-disorder transition*.

A high temperature example of a specific heat anomaly is to be found in β-brass, the 50/50 copper–zinc alloy. At about 460°C there is a large peak in the specific heat (Fig. 5.4), indicating a local change of order. Subtracting the background one obtains the anomalous contribution to the specific heat, c'. Integration of c'

[5] H. Moser, *Phys. Zeit.*, **37**, 737, 1936.

yields the increase in energy and integration of c'/T yields the increase in entropy associated with the change of order. For 1 mol the entropy change is close to $N_A k \ln 2 = 5{\cdot}8$ J K^{-1} suggesting a two-fold change in order per pair of copper and zinc atoms. The explanation which has been confirmed for similar alloys by X-ray studies,[6] is that in the low temperature form the copper and zinc atoms are arranged in a regular array whereas in the high temperature form they are randomly distributed on the lattice sites. The crystal structure of β-brass is body-centred cubic so that the ordered array corresponds to, say, copper atoms at the corners of the cubes and zinc atoms at the centres. The probability, in the disordered structure, of finding a particular kind of atom at a particular lattice site is clearly $\frac{1}{2}$ since there are equal numbers of copper and zinc atoms; whence we have for the change of entropy:

$$\Delta S = k \ln \frac{g_{\text{disordered}}}{g_{\text{ordered}}} = k \ln 2^{N_A} = R \ln 2.$$

Another example of an order-disorder transition is to be found in the low temperature behaviour of paramagnetic salts. Here, the specific heat peak, which is usually within a few degrees of absolute zero, results from a change in the magnetic order. Paramagnetism is always associated with the presence of microscopic magnetic dipoles which may be aligned by an external field to produce a net total magnetization. Even in the absence of an applied field, however, the different orientations possible for the dipoles have slightly different energies as a result of their interactions with one another and with the crystal lattice in which they are situated. At low temperatures, they will all occupy the lowest available levels and the material will be magnetically ordered. In many cases, the ordered state corresponds to the parallel alignment of ferromagnetism with its large net magnetization, but this is not the only form of ordering which occurs.[7] As the temperature is raised, the dipoles become excited into the higher levels, and at high temperatures, they become randomly distributed among the orientations. Thus, over the range of temperatures where the ordering sets in, there is an

[6] The experiment cannot be done on β-brass. The difference between the atomic weights of copper and zinc is so small that their scattering power for X-rays is very similar and it is not possible to distinguish the two kinds of atom.

[7] See, C. Kittel, *Introduction to Solid State Physics*, Chapter 16, Wylie, 1971.

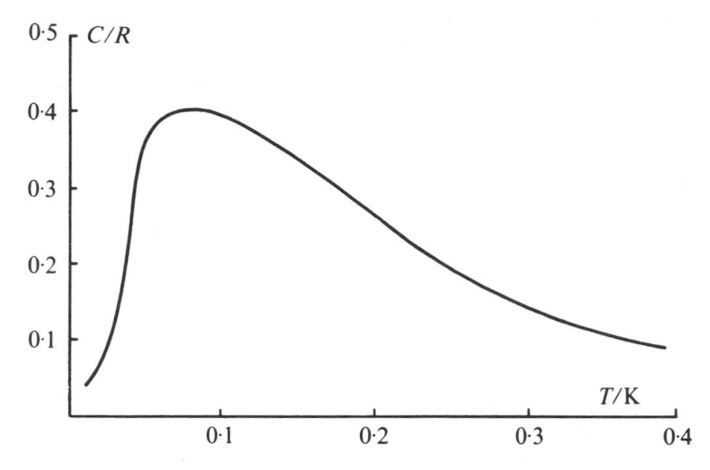

Fig. 5 . 5. The low temperature heat capacity of chromium potassium alum.[8]

extra contribution to the heat capacity deriving from the change in magnetic order. The anomaly in chromium potassium alum is shown in Fig. 5.5. Here there are four possible orientations for the dipoles,[9] so that the probability of finding a particular dipole in a specified orientation is $\frac{1}{4}$. The change in entropy per element in proceeding from the ordered to the disordered state is therefore $k \ln 4$, and the entropy change for one mole, $k \ln 4^{N_A} = R \ln 4 = 1{\cdot}39\, R$. Measurements to the lowest temperatures indicate that the entropy change associated with the ordering is indeed of this order.[10]

It should be noted that in the above examples the magnitudes of the changes in entropy are similar, but the transitions occur at very different temperatures. The entropy change, of course, is simply dependent on the change of order and plays no part in determining the transition temperature. The latter is determined by energy considerations, being that temperature at which the thermal energy becomes comparable with the energy associated with the ordering process. This gives for the transition temperature,

[8] B. Bleaney, *Proc. Roy. Soc.*, **A204**, 216, 1950; and D. de Klerk, M. J. Steenland and C. J. Gorter, *Physica*, **15**, 649, 1949.

[9] The magnetic moment is associated with the Cr^{+++} ion, which at low temperatures behaves as if it were in a ^4S state with $J = S = \frac{3}{2}$ and $g = 2$. See, D. de Klerk, 'Adiabatic Demagnetisation', *Encyclopaedia of Physics*, XV, Springer-Verlag, 1956.

[10] D. de Klerk, M. J. Steenland, and C. J. Gorter, *Physica*, **15**, 649, 1949.

$T \sim \varepsilon/k$, where ε is the energy required to remove one element from the ordered state. (That is, to exchange a copper atom with a zinc atom in the ordered brass or to disalign a dipole in the ordered paramagnetic.) The corresponding energies in the examples above are about 10^{-20} J $\simeq 60$ meV and 10^{-24} J $\simeq 7$ μeV.[11]

5.6.3. Latent Heats

In the examples we have discussed above, we have associated specific heats with the gradual change of entropy associated with the gradual change of order as the temperature changes. Latent heats correspond to a sudden change in order associated with a first order[12] phase change such as the melting of a solid or vaporization of a liquid (Fig. 5.6).

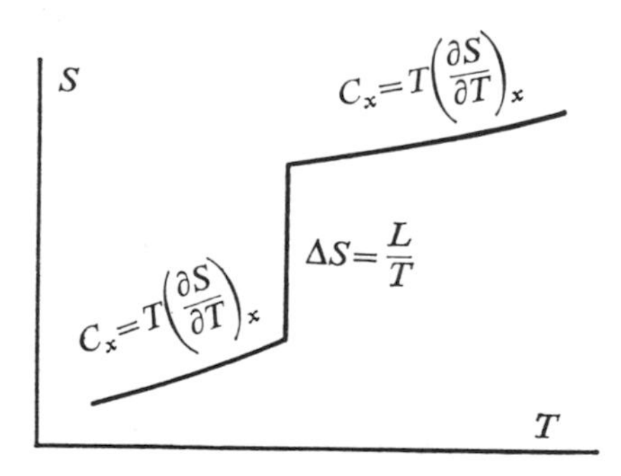

Fig. 5 . 6. Entropy near a first order change of phase.

We may make a very crude estimate of the entropy change associated with vaporization. If we think of the molecules in the liquid as moving about freely, in a gas-like manner, but as being restricted to a much smaller volume than when they are in the vapour phase, then the ratio of the statistical probabilities of finding any one molecule in the large volume available in the vapour rather than in the small volume available in the liquid is simply equal to the ratio of the available volumes. (Cf. the illustration at the beginning of section 5.6 where the volumes are equal.) Thus, the

[11] It is often convenient to express energies of atomic-sized systems in electronvolts. It is useful to remember that $k = 1{\cdot}38 \times 10^{-23}$ J $= 86{\cdot}3$ μeV.

[12] Change of phase will be discussed in detail in chapter 10.

ratio of the statistical probabilities of the vapour and liquid configurations for all N_A molecules of one mole is

$$\frac{g_{\text{vapour}}}{g_{\text{liquid}}}=\left(\frac{V_{\text{vapour}}}{V_{\text{liquid}}}\right)^{N_A}=\left(\frac{\rho_{\text{liquid}}}{\rho_{\text{vapour}}}\right)^{N_A}.$$

For many substances, the density ratio is about 10^3. This gives the entropy change associated with vaporization as

$$\Delta S=k \ln 10^{3N_A}=R \ln 10^3 \sim 7R.$$

Relating this to the latent heat, we have

$$\frac{L}{T_b R}\sim 7,$$

where T_b is the boiling point.
This corresponds to Trouton's rule, found empirically, which states that for non-associated liquids,

$$\frac{L}{T_b R}\sim 10.$$

The restriction to non-associated liquids is necessary, because when association takes place a new degree of order is introduced in the liquid giving rise to a further contribution to the entropy change.

5.6.4. Change in Order by Deformation

In the three examples above, change in order was brought about by changing the temperature of the system. Change in order is also generally brought about when work is done on a system under conditions where heat may be exchanged with the surroundings. (If heat could not be exchanged, the entropy would be invariant under reversible changes, so that the statistical order would also be invariant. Irreversible work, of course, always brings about an increase in entropy and a decrease in order.) An illuminating example is to be found by contrasting the effects of mechanical deformation in different kinds of solid.

If a metal wire is stretched adiabatically, it cools. If rubber is stretched adiabatically the temperature rises. This is easily understood in terms of their microscopic properties.

The metal of the wire consists of many small crystallites in each of which the atoms are arranged in a regular lattice. When the wire is stretched, each crystallite is distorted and loses some of its symmetry. (For example, the lattice might be distorted from cubic to tetragonal.) The loss in symmetry is a loss of order and corresponds to an increase in entropy. If the distortion were performed under isothermal conditions, heat would therefore be absorbed in order that the entropy increase. When performed under adiabatic conditions the total entropy is kept constant by a drop in temperature:

$$\left(\frac{\partial T}{\partial x}\right)_S = -\left(\frac{\partial T}{\partial S}\right)_x \left(\frac{\partial S}{\partial x}\right)_T.$$

(The first term on the right is related to a principal specific heat, and is always positive.)

The molecular arrangement in rubber, on the other hand, is very different from that of a crystal. Rubber consists of long organic molecules which are normally tangled together in a random manner. When the rubber is stretched these long molecules tend to align along the direction of extension and the order increases. Therefore, in this case, when the material is stretched isentropically the temperature rises.

6. The Carathéodory Formulation of the Second Law

6.1. Introductory Remarks

In the last two chapters, we have stated and developed the second law of thermodynamics along traditional lines. The statements of the law asserted the impossibility of certain processes which are easily visualized and readily believed. However, to arrive at the real substance of the matter, we had to equip ourselves with the paraphernalia of idealized heat engines and wade through lengthy arguments about efficiencies and cyclic processes. Only then did we discover that we had arrived, as if by good fortune, at a new function of state, the entropy, on which depends all the subsequent development of the subject. In fact, the essential function of the second law is to enable us to define this quantity and to derive its properties. It seems desirable, therefore, to adopt a formulation of the law which achieves this end with greater economy. That put forward early this century by Carathéodory[1] does precisely that.

One may well enquire why, if it has these advantages, Carathéodory's statement of the second law is not more widely used. There are two reasons for this. In the first place, any formulation which makes it possible to avoid the use of cycles and heat engines in the basic development must necessarily be framed in somewhat more abstract terms than the Kelvin or Clausius statements which refer to specific processes. This, it is argued, makes it less easy to assimilate. There is some truth in this; and that is why, in this book, we have first developed the subject along traditional lines; but now, having gained some insight into entropy and its properties, we may return and, with very little effort, replace the devious treatment by the more direct.

[1] C. Carathéodory, *Math. Ann.* **67**, 355, 1909; and *Sitz. d. Preu. Akad. d. Wiss.* p. 39, 1925.

The second, and perhaps more significant reason why the older forms of the second law have not been discarded is that Carathéodory's original development was set in very mathematical terms requiring of most scientists such an effort that the physical simplicity of his idea became obscured. Recently, there has been considerable discussion of his formulation, and, due to the efforts of H. A. Buchdahl and others, it has been stripped of the greater part of its mathematics. The exposition we give here derives from that proposed by Buchdahl [2].

Any who wish may omit this chapter, for we have already developed what is necessary for the rest of the book. However, those who do so will miss an opportunity of gaining greater insight into the meaning of the second law, for the Carathéodory formulation brings out its essential features in a way which the traditional treatment does not.

6.2. Empirical Entropy

Carathéodory's statement of the second law, often referred to as *Carathéodory's Principle*, reads:

In the neighbourhood of any arbitrary state J of a thermally isolated system Σ, there are states J' which are inaccessible from J. A state is said to be in the neighbourhood of J if its state variables differ from those of J by however small an amount. The Carathéodory statement, therefore, asserts that we may find states *as close as we please* to J which are inaccessible in an adiathermal change. In the light of what we already know about the second law, the states which are inaccessible from J by adiathermal processes are those whose entropy is less than that of J. We must now show that we may deduce the existence and derive the properties of entropy from the Carathéodory statement.

For convenience, we introduce the following notation:
For our thermally isolated system,

$J \not\rightarrow J'$ means that J' is inaccessible from J,

$J \rightarrow J'$ means that J' is accessible from J but not the reverse,

$J \leftrightarrow J'$ means that J' is accessible from J and vice versa.

We shall describe the state of Σ by choosing for it a set of independent variables. For these it is convenient to take the extensive

[2] H. A. Buchdahl, *Z. Phys.* **152**, 425, 1958 (in English), and *The Concepts of Classical Thermodynamics*, Cambridge University Press, 1966. In earlier papers he gives simplified versions of the original Carathéodory treatment. See: *Amer. J. Phys.* **17**; 41, 44, 212, 1949.

variables associated with all work-like processes to which the system is subject, (for example, volume, surface area, magnetization, etc.), and one other convenient function of state.[3] The extensive variables are called the *configuration coordinates* of the system since they define its configuration (volume, surface area, etc.) but not, of course, its state for which one further piece of information is necessary.

We first pause to show that it is impossible for two states to be mutually inaccessible. We do this by the following argument: Consider any two states J_1 and J_2 of a system Σ. Take as the independent system parameters, the configuration coordinates x_i and the internal energy U. Suppose, first, that we start from J_1 and reversibly deform the system until the x_i take the values appropriate to the state J_2, and suppose that in doing this the internal energy becomes U' corresponding to a state J' which will, in general, be different from J_2.[4] Then we will have either $U' < U_2$, or $U' = U_2$, or $U' > U_2$. If $U' < U_2$ we may, keeping the x_i fixed, increase U, by some irreversible process such as stirring, to U_2. (We know from experience that we can *increase* U by irreversible work. We do not yet know that we cannot decrease it. That follows from the second law.[5]) We have then brought Σ to J_2 by a particular adiathermal path, so that $J_1 \rightarrow J_2$. If $U' = U_2$, then $J' = J_2$: We have reached J_2 immediately by a path which is reversible and $J_1 \leftrightarrow J_2$. If $U' > U_2$, then we may start from J_2 and first increase U from U_2 to U' (by stirring) keeping the x_i constant. Σ is then in the state J' and we may bring it to J_1 by the same reversible path as before. In this case, $J_2 \rightarrow J_1$. We have thus proved, using only the first law, that

[3] For a system with n degrees of freedom there are $(n-1)$ conjugate pairs of variables like (p, V), (B, m), each associated with a process by which *work* may be performed on the system. The remaining degree of freedom corresponds, of course, to the pair (T, S) associated with processes involving *heat*. However, in the present context, we have not yet shown that these quantities exist and so are forced to choose the remaining variable from elsewhere.

[4] It is not necessary to this argument that J' be uniquely defined. At the moment, we have no justification for expecting it to be, for to fix the state of the system not only the x_i but also U must be given. We are thus not deforming Σ to a specified *state* but to a specified *configuration* (given x_i) for which U' may not be uniquely determined. (It might depend on path.) In fact, of course, it *is* uniquely determined, for the deformation is isentropic and this constraint suffices to define the final state. However, this statement follows from the second law through arguments we have not yet made.

[5] This observation has the status of a subsidiary axiom.

two states of a thermally isolated system cannot be mutually inaccessible. The symbols $\not\leftrightarrow$ and $\leftarrow$ are thus entirely equivalent. We now return to the main line of argument.

Consider the two states J_1 and J_2 and allot to these states numbers σ_1 and σ_2 which are arbitrary except for the restriction that

$$\begin{aligned} \sigma_1 > \sigma_2 \quad &\text{if} \quad J_1 \not\leftrightarrow J_2 \\ \sigma_1 = \sigma_2 \quad &\text{if} \quad J_1 \leftrightarrow J_2 \\ \sigma_1 < \sigma_2 \quad &\text{if} \quad J_1 \leftarrow\!\!\!\!/\!\!\!\!\rightarrow J_2. \end{aligned}$$

These numbers have the characteristics of 'thermodynamic weights' for the states J_1 and J_2. As we might expect, we shall eventually extract from them the quantity which will be the entropy.

Now consider another state J_3 to which we allot the number σ_3 which is again arbitrary except that we require that

$$\begin{aligned} &\text{if} \quad J_3 \not\leftrightarrow J_1, \quad \sigma_3 > \sigma_1, \\ &\text{if} \quad J_3 \leftrightarrow J_1, \quad \sigma_3 = \sigma_1, \\ &\text{if} \quad J_3 \leftarrow\!\!\!\!/\!\!\!\!\rightarrow J_1, \quad \sigma_3 < \sigma_1, \end{aligned}$$

etc., for J_2.

This procedure is internally consistent. For example, the statement 'if $\sigma_1 > \sigma_2$ and $\sigma_2 > \sigma_3$ then $\sigma_1 > \sigma_3$' implies 'if $J_1 \not\leftrightarrow J_2$ and $J_2 \not\leftrightarrow J_3$ then $J_1 \not\leftrightarrow J_3$' which must necessarily be the case, for if, instead, $\sigma_1 \leq \sigma_3$ were possible, then Σ could make the transition from J_1 to J_3 and thence back to J_2 (since $J_2 \not\leftrightarrow J_3$ implies $J_2 \leftarrow J_3$) thus effecting a transition from J_1 to J_2 which contradicts $\sigma_1 > \sigma_2$.

It is thus possible, in a self consistent manner, to allot *all* states of Σ numbers, such that for any two states, $\sigma_1 > \sigma_2$, $\sigma_1 = \sigma_2$, or $\sigma_1 < \sigma_2$ according as $J_1 \not\leftrightarrow J_2$, $J_1 \leftrightarrow J_2$, or $J_1 \leftarrow\!\!\!\!/\!\!\!\!\rightarrow J_2$. This set of numbers performs a function in relation to adiathermal accessibility rather like that of empirical temperature in relation to thermal equilibrium. (Section 2.2.) Just as equality of temperature implied thermal equilibrium, so equality of σ implies mutual accessibility under thermal isolation. Because of this similarity, we may call σ the *empirical entropy*. The difference, however, is that here we attach a special meaning to the ordering of the numbers.

It is worth pointing out that thus far we have implicitly used Carathéodory's statement of the second law in the restricted form that inaccessible states do exist. We apply it fully in the argument that follows.

Now it is possible to allot the numbers for empirical entropy in such a way as to be a single-valued continuous function of the system parameters.[6] We may show this by demonstrating one particular system for allotting them. Suppose we again describe the system Σ by its configuration coordinates x_i and the internal energy U. Then we may arbitrarily choose a reference state J_0 for which we put $\sigma = \sigma_0$. To define σ for another state J_1, we start from J_1 and deform the system reversibly until the x_i take the values appropriate to the state J_0. In general, Σ will not then be in the state J_0 with internal energy U_0, but in some other state J' with internal energy U'. We first show that J' is unique. That is, that the *state* after the deformation does not depend on path.[7]

Suppose that by reversible deformation from J_1 to the *configuration* corresponding to the given x_i it were possible to arrive at two different states, J' with internal energy U' or J'' with internal energy U''. Suppose that $U'' > U'$. Then it is possible to pass from J'' to *any* state in its neighbourhood by the following route:

(*a*) Pass reversibly from J'' to J_1 and thence to J'.

(*b*) Change the configuration coordinates, the x_i (which are the same as for the state J'') to values corresponding to the *configuration* of any state in the neighbourhood of J''. Since this change may be arbitrarily small, the internal energy after it, U''', even though greater than U' may always be made less than U''.

(*c*) We may now increase U from U''' to any desired value in the region of U'' by some irreversible process (such as stirring) thus reaching any desired *state* in the region of J''.

The possibility of doing this violates the second law and we therefore conclude that U' and J' are uniquely determined.

This being so, we are able to define a satisfactory scale of empirical entropy by taking σ at J_1 to be given by

$$\sigma_1 = \sigma_0 + (U' - U_0) \tag{6.1}$$

where $(U' - U_0)$ may be measured by some suitable experiment. We have already shown that this defines a single-valued scale and its continuity follows from the continuity of the energy function. Furthermore, we see that it has the properties we require of empirical entropy in that

(*a*) if $U' = U_0$ then J_1 and J_0 are mutually accessible by the reversible path.

(*b*) if $U' > U_0$ then J_1 is accessible from J_0 but not the reverse, and

[6] This is what we did for empirical temperature in defining a temperature scale.

[7] This deals with the point we raised in footnote (4).

(*c*) if $U' < U_0$ then J_0 is accessible from J_1 but not the reverse.

We have thus established that there is a continuous single-valued function of state with the properties of empirical entropy. From it we may clearly generate others by putting $\sigma' = f(\sigma)$ where f is any monotonically increasing function of σ.

The existence and uniqueness of reversible adiabatics (for systems of any degree of complexity) follows immediately, since these are simply the surfaces for which $\sigma =$ constant.

6.3. Empirical Entropy and Heat

So far, we have confined ourselves to discussing thermally isolated systems where we have been particularly concerned with the idea of inaccessibility. We must now relate empirical entropy to processes in systems which are not thermally isolated.

It is now convenient to choose as our independent system variables the configuration coordinates x_i and the temperature Θ measured on *any empirical scale*. Then we may adopt any suitable prescription, such as that provided by equation (6.1), for defining a scale of empirical entropy as a single-valued, continuous function of the system variables:

$$\sigma = \sigma(x_i, \Theta) \tag{6.2}$$

which may be differentiated to give $d\sigma$ as a sum of terms linear in the differentials of the system variables:

$$d\sigma = \Sigma\, \xi_i\, dx_i + \zeta\, d\Theta. \tag{6.3}$$

Now, for an infinitesimal reversible change the first law states

$$đQ_{rev} = dU - \Sigma\, X_i\, dx_i,$$

which, since U is a function of the x_i and Θ, may be rearranged in a form similar to that of (6.3), namely

$$đQ_{rev} = \Sigma\, \xi_i'\, dx_i + \zeta'\, d\Theta. \tag{6.4}$$

Now for a reversible *adiabatic* change we have

$$đQ_{rev} = 0,$$

and, by the properties of empirical entropy,

$$d\sigma = 0.$$

But in the absence of the adiabatic constraint all the terms on the right of equations (6.3) and (6.4) are independent. Therefore, the only way in which $d\sigma = 0$ can be satisfied whenever $đQ_{rev} = 0$ is for $d\sigma$ and $đQ_{rev}$ to be in a simple relationship of the form

$$đQ_{rev} = \lambda \, d\sigma \tag{6.5}$$

where λ is a non-zero function of the variables of state.

Now σ is a function of state, and therefore $d\sigma$ is necessarily integrable. From (6.5) it follows that $đQ_{rev}/\lambda$ is also a perfect differential. We have thus shown that there is always an integrating factor for $đQ_{rev}$, although, so far, all we know about it is that it is some function of state.

6.4. Thermodynamic Temperature and Entropy

We now show that $\lambda d\sigma$ may be put in the form $T\,dS$, where T is a *universal* function of temperature only.

To do this we consider two systems Σ' and Σ'' which are in thermal contact and at equilibrium. As independent variables for each of these we now take (because it is convenient to do so) the temperature on some empirical scale, Θ, the empirical entropy σ, and $(n-2)$ of the other variables x_i, where n is the number of degrees of freedom of the system. Thus the state of Σ' is defined by (x_i', σ', Θ) and of Σ'' by $(x_i'', \sigma'', \Theta)$. We may also consider the two systems taken together as making up a single composite system Σ. To specify the state of Σ we require all the independent variables of both subsystems, namely $(x_i', x_i'', \sigma', \sigma'', \Theta)$. Because of thermal equilibrium, Θ is, of course, the same for Σ', Σ'', and Σ.

If, now, we add heat reversibly to the composite system, we have, by the first law

$$đQ_{rev} = đQ'_{rev} + đQ''_{rev} \tag{6.6}$$

which, by (6.5), gives

$$\lambda \, d\sigma = \lambda' \, d\sigma' + \lambda'' \, d\sigma''$$

or

$$d\sigma = \frac{\lambda'}{\lambda} \, d\sigma' + \frac{\lambda''}{\lambda} \, d\sigma''. \tag{6.7}$$

Now we have shown that empirical entropy is a function of state so that $d\sigma$ is a perfect differential. Then equation (6.7) shows that σ may be expressed as a function of σ' and σ'' only. Therefore we must also have:

$$\frac{\lambda'}{\lambda} \text{ and } \frac{\lambda''}{\lambda} \text{ are functions of } \sigma' \text{ and } \sigma'' \text{ only.} \tag{6.8}$$

But, *a priori*, we have expected the λ's to contain *all* the appropriate state variables; namely,

$$\lambda' = \lambda' \ (x_i', \sigma', \Theta)$$
$$\lambda'' = \lambda'' \ (x_i'', \sigma'', \Theta)$$
$$\lambda = \lambda \ (x_i', x_i'', \sigma', \sigma'', \Theta).$$

Then, applying (6.8), if

$$\frac{\lambda'(x_i', \sigma', \Theta)}{\lambda(x_i', x_i'', \sigma', \sigma'', \Theta)} \text{ depends only on } \sigma' \text{ and } \sigma'',$$

then, certainly, λ cannot contain the x_i''. It could, of course still contain the x_i' in such a way that they cancelled out of the ratio λ'/λ. However, if we apply (6.8) to the ratio λ''/λ we see that λ cannot contain the x_i' either. Thus

$$\lambda = \lambda(\sigma', \sigma'', \Theta). \tag{6.9}$$

Returning to (6.8) in the light of (6.9) it follows that

$$\lambda' = \lambda'(\sigma', \Theta)$$

and

$$\lambda'' = \lambda''(\sigma'', \Theta).$$

For (6.8) to be true now, we only require that the Θ dependence of the numerators and denominators of the ratios be the same so that it cancels out. Thus, in fact, each of the λ's must factorize into the form

$$\lambda(\sigma, \Theta) = T(\Theta) f(\sigma) \tag{6.10}$$

where $T(\Theta)$ is a *universal* function of the empirical temperature. (It is the same for all systems at the same temperature.)
Equation (6.5) now becomes

$$đQ_{\text{rev}} = T(\Theta) f(\sigma)\, d\sigma.$$

But we may put

$$dS = f(\sigma)\, d\sigma$$

where S will also be a function of state (but for an arbitrary constant). This gives $đQ_{\text{rev}}$ in the form we require

▶ $$đQ_{\text{rev}} = T\, dS \tag{6.11}$$

where T is *thermodynamic temperature* and S *entropy*.[8] Substituting back in (6.6),

$$dS = dS' + dS'' = d(S' + S'')$$

which shows that entropy changes are additive. Integrating,

$$S = S' + S''$$

where we have *chosen* to put the integration constant equal to zero, emphasizing that in the context of the second law only entropy *changes* have physical significance. Entropy is thus an *extensive* variable and thermodynamic temperature its conjugate *intensive* variable.

6.5. Irreversible Changes

We may now arrange for S always to increase in irreversible changes of a thermally isolated system by adopting an appropriate sign convention. We first show, by considering a particular irreversible change, that λ is necessarily positive. Consider an infinitesimal irreversible change in a thermally isolated system which we bring about by doing irreversible work in such a way that the configuration coordinates at the end are the same as they were at the beginning. We know that this can increase the internal energy[5]: $dU > 0$, and since the change is irreversible, $d\sigma > 0$. But we may also take the system from the initial to the final state reversibly if we remove the adiabatic constraint. This requires no work (since the configuration coordinates do not change) but only addition of heat $đQ_{rev}$. But $đQ_{rev} = dU > 0$. Hence, applying $đQ_{rev} > 0$ and $d\sigma > 0$ to (6.5) we see that λ is necessarily positive. From (6.10) we therefore see that we may choose either that both T and f are positive or that they are both negative. We choose the positive sign. Then by the relation $dS = f(\sigma)d\sigma$ we see that $d\sigma > 0$ gives $dS > 0$. But the essence of an irreversible change is that the original state becomes inaccessible and $d\sigma > 0$. Therefore, for any change in a thermally isolated system we must have

$$dS \geq 0 \tag{6.12}$$

where the equality applies only to reversible changes.

[8] Note that only *changes* in entropy are defined by the second law. In the context of classical thermodynamics no significance can be attached to its absolute value. (See section 5.6.)

For a system which exchanges heat with its surroundings, we simply consider the system and the surroundings as a composite system. Then by (6.12),

$$\mathrm{d}S_{\text{total}} = \mathrm{d}S + \mathrm{d}S_0 \geq 0 \tag{6.13}$$

where $\mathrm{d}S$ refers to the system and $\mathrm{d}S_0$ to the surroundings. If any irreversibility occurs in the system only, we may put

$$đQ = -T_0 \, \mathrm{d}S_0$$

where $đQ$ is the heat supplied *to* the system and we have the result

▶ $$đQ \leq T_0 \, \mathrm{d}S \tag{6.14}$$

where T_0 is the temperature *of the source of heat* and $\mathrm{d}S$ is the change in entropy *of the system*. This is the fundamental result of section 5.2.

For a reversible change, which requires a system to be in thermal equilibrium with a source of heat when heat is exchanged, $T = T_0$ in (6.14), the equality holds, and the first law becomes

$$\mathrm{d}U = T \, \mathrm{d}S + \Sigma \, X_i \, \mathrm{d}x_i. \tag{6.15}$$

This completes the derivation of the fundamental consequences of the second law; but before we see how the subsequent development continues, it is instructive to pause and examine where the element of choice has entered into the treatment and how the choice is reflected in the properties of entropy and temperature.

The Carathéodory statement asserts the presence of states which are inaccessible to a thermally isolated system. From this we deduce the existence of a function of state which is constant in reversible changes but changes *in one direction or the other* in an irreversible change. We *chose* to define it as being unable to decrease. Later we introduced the non-zero function of state, λ, and showed, by referring to an irreversible process which we know to occur, that it is necessarily positive. We then *chose* to have S increase in irreversible changes which required $f(\sigma)$ and hence T to be positive. A further property of absolute temperature now follows. To see what this is we refer to the simple process involving irreversibility which we have already discussed in the context of the first law in defining the terms hotter and colder. (Section 3.4.)

If we connect two bodies at different temperatures by a thermal path, heat will flow from the hotter to the colder body. When a small amount of heat $đQ$ (a positive quantity) flows from one to the other, the associated change in the total entropy is

$$\mathrm{d}S = \mathrm{d}S_1 + \mathrm{d}S_2 = đQ \left(\frac{1}{T_2} - \frac{1}{T_1} \right) \tag{6.16}$$

where T_1 is the temperature of the hotter body and T_2 that of the colder. Then we have chosen that $\mathrm{d}S > 0$, whence

$$T_2 < T_1. \tag{6.17}$$

Therefore it is a consequence of our choice that a hotter body has a larger value of T than a colder one.

The 1 : 1 correspondence between hotness and temperature now follows immediately from (6.16) since if bodies at one temperature could be either hotter or colder than those at another, it would be possible to choose a pair for which heat would flow from the lower temperature (smaller T) to the higher (larger T) causing a spontaneous decrease of entropy in violation of the principle of increase of entropy.

6.6. Subsequent Development

In the previous sections, we derived all the fundamental results of the last two chapters without recourse to engines and cyclic processes. If we now turn to such practical matters as the interconversion of heat and work we may quickly derive the appropriate results for heat engines and refrigerators.

To analyse the conversion of heat into work we distinguish the sources of heat and work from the device which supplies the mechanism to effect the conversion. We must suppose that in the whole process the device is unchanged, and this is why we introduced the idea of cyclic processes in chapter 3. Consider then the effect of an infinitesimal change in which a heat engine absorbs heat $đQ$ from a reservoir at temperature T_0. Then the work done *by* the heat engine in the change is, by the first law,

$$đW = đQ - \mathrm{d}U$$

or, by (6.14),

$$đW \leq T_0\,\mathrm{d}S - \mathrm{d}U. \tag{6.18}$$

The work is therefore a maximum when the equality holds which is the case for a reversible change. This requires thermal equilibrium between the engine and the source of heat : $T = T_0$.
Then (6.18) becomes

$$đW = T\,\mathrm{d}S - \mathrm{d}U. \tag{6.19}$$

After a complete cycle, the engine returns to its initial state. Integrating around the cycle

$$\oint \mathrm{d}U = 0 \tag{6.20}$$

because U is a function of state. Hence, from (6.19), the total work done in one cycle is

$$W = \oint T \, dS. \tag{6.21}$$

But we also have

$$\oint dS = 0 \tag{6.22}$$

since S is a function of state. It is clear that (6.22) may only be satisfied if heat is both absorbed *and rejected* in different parts of the cycle. It follows immediately that it is impossible to devise a process which will convert heat entirely into work. (The Kelvin statement of the second law.)

The simplest way in which (6.22) may be satisfied is for the engine to exchange heat at two reservoirs only. The corresponding reversible cycle is simply the Carnot cycle for which (6.22) and (6.21) give

$$\frac{Q_1}{Q_2} = \frac{T_1}{T_2} \tag{6.23}$$

and

$$W = Q_1 \left(1 - \frac{T_2}{T_1}\right) \tag{6.24}$$

where T_1 and T_2 are the temperatures of the reservoirs. These are the results of section 4.6.

For a refrigerator or heat pump we wish to minimize the work required to extract a given amount of heat from the cold reservoir. The work done *on* the system in an infinitesimal change is

$$\begin{aligned} đW &= dU - đQ \\ &\geq dU - T_0 \, dS. \end{aligned} \tag{6.25}$$

This is minimized when the equality holds and the change is reversible. We then again have (6.23) for the heat exchanged at the two reservoirs, which gives

$$\frac{Q_1}{W} = \frac{T_1}{T_1 - T_2} \tag{6.26}$$

and

$$\frac{Q_2}{W} = \frac{T_2}{T_1 - T_2} \tag{6.27}$$

which are the results of section 4.8.

7. Thermodynamic Potentials

7.1. The Potential Functions

We have already, in the context of the first law, defined two functions of state with the dimensions of energy: the internal energy and the enthalpy. Clearly, we may invent others by adding to the internal energy or the enthalpy any other function of state with the dimensions of energy. Few of these would have any particular physical significance, but some of them do have, and play an important part in thermodynamics. Because of their role in determining the equilibrium states of systems under various constraints (section 10.2), they are known as the *thermodynamic potential functions*.

For a system with two degrees of freedom, there are four thermodynamic potentials. Referring again to a system subject to work by hydrostatic pressure only, they are[1]

$$\left.\begin{array}{lll} \blacktriangleright & \text{internal energy,} & U \\ \blacktriangleright & \text{enthalpy,} & H = U + pV \\ \blacktriangleright & \text{Helmholtz function,} & F = U - TS \\ \blacktriangleright & \text{Gibbs function,} & G = U - TS + pV. \end{array}\right\} \quad (7.1)$$

The analogous functions where the work is not done by hydrostatic pressure are obtained by replacing $-p$ and V by the appropriate pair of variables (γ, A; B, m; etc. See Table 3.1).

The significance of the potential functions becomes a little more apparent from their differential forms which follow by differentiating

[1] These are the names and symbols usually used by physicists. Other common variants are: Internal energy is sometimes given the symbol E. Enthalpy is sometimes called the *heat content*. (See section 3.7.) The Helmholtz function is sometimes called the (Helmholtz) *free energy* or the *work function* and is sometimes given the symbol A. The Gibbs function is sometimes called the *free energy* or the *thermodynamic potential* and given the symbol F!

the equations of (7.1) and using the result we have already derived for $\mathrm{d}U$ (section 5.4):

$$\left.\begin{array}{lll} \blacktriangleright & \text{internal energy,} & \mathrm{d}U = T\,\mathrm{d}S - p\,\mathrm{d}V \\ \blacktriangleright & \text{enthalpy,} & \mathrm{d}H = T\,\mathrm{d}S + V\,\mathrm{d}p \\ \blacktriangleright & \text{Helmholtz' function,} & \mathrm{d}F = -S\,\mathrm{d}T - p\,\mathrm{d}V \\ \blacktriangleright & \text{Gibbs' function,} & \mathrm{d}G = -S\,\mathrm{d}T + V\,\mathrm{d}p. \end{array}\right\} \quad (7.2)$$

Each has two terms on the right corresponding to the two degrees of freedom of the system. These terms derive from the two pairs of fundamental variables which appear in the expression for $\mathrm{d}U$; namely, (T, S) and (p, V).

Because the potentials are functions of state, their differentials are exact. Then the equations of (7.2) show that each potential has a different pair of the fundamental variables as its natural or *proper* independent variables: $U=U(S, V)$, $H=H(S,p)$, $F=F(T, V)$, and $G=G(T, p)$.

If any one of the potentials is known *explicitly* in terms of its proper variables then we have complete information about the system, for we may calculate any of the parameters of state from the one function. As an example, we will consider the Helmholtz function, F, which is particularly important in connection with statistical mechanics since the expression for F in terms of statistical parameters is very simple and forms a link between the microscopic analysis and the macroscopic variables.

The proper variables for F are T and V. If F is given explicitly in terms of these we see from the differential form, (7.2), that the other two fundamental variables, S and p, follow immediately:

$$S = -\left(\frac{\partial F}{\partial T}\right)_V \quad \text{and} \quad p = -\left(\frac{\partial F}{\partial V}\right)_T.$$

The expressions for U, H, and G may then be constructed directly from their definitions:

$$F = U - TS,$$

so

$$U = F + TS = F - T\left(\frac{\partial F}{\partial T}\right)_V = -T^2 \left.\frac{\partial}{\partial T}\right)_V \left(\frac{F}{T}\right).$$

This expression for U in terms of F is known as the *Gibbs–Helmholtz equation*. Similarly,

$$H = U + pV = F - T\left(\frac{\partial F}{\partial T}\right)_V - V\left(\frac{\partial F}{\partial V}\right)_T$$

and
$$G = F + pV = F - V\left(\frac{\partial F}{\partial V}\right)_T = -V^2\left(\frac{\partial}{\partial V}\right)_T\left(\frac{F}{V}\right).$$

It should be pointed out that, as with the other functions of state, given suitable information it is always possible to calculate how a potential function changes when the system goes from one state to another. For instance, if we know $G(T_0, p_0)$ but wish to calculate $G(T_0, p_1)$ we may write

$$G(T_0, p_1) - G(T_0, p_0) = \int_{p_0}^{p_1}\left(\frac{\partial G}{\partial p}\right)_T \mathrm{d}p.$$

From the differential form of G we see that this becomes:

$$G(T_0, p_1) - G(T_0, p_0) = \int_{p_0}^{p_1} V\,\mathrm{d}p.$$

To evaluate this the only information required is V as a function of p.

Some of the more important properties of the potential functions for a system subject to work by hydrostatic pressure only, may be summarized as follows:

Internal energy. For a thermally isolated system, $\mathrm{d}U = \text{đ}W$ and the decrease in the internal energy is equal to the work done *by* the system. If the change of state takes place isentropically (reversibly as well as under conditions of thermal isolation), then the work done by the system is $-\Delta U = p\,\mathrm{d}V$. For an isovolumic change, $\mathrm{d}V = 0$ and $\text{đ}W = 0$ so that the change in the internal energy is the heat absorbed: $\mathrm{d}U = \text{đ}Q_V$. Hence,

$$C_V = T\left(\frac{\partial S}{\partial T}\right)_V = \left(\frac{\partial U}{\partial T}\right)_V.$$

Enthalpy. In an isentropic change (a reversible change in a thermally isolated system) the change in the enthalpy is related to the change in the pressure. In a reversible isobaric process, the change in enthalpy is the heat entering the system: $\mathrm{d}H = \text{đ}Q_p$. Hence

$$C_p = T\left(\frac{\partial S}{\partial T}\right)_p = \left(\frac{\partial H}{\partial T}\right)_p.$$

We have also shown that the enthalpy is the total energy transported internally (i.e., not including kinetic or potential forms) by a flowing fluid. (Section 3.8.)

Helmholtz function. In an isothermal change the decrease in the Helmholtz function is the maximum amount of mechanical work which may be extracted from the system. (Hence the alternative name: Helmholtz *free* energy.) If the change were irreversible, the work done would be smaller than $p\,\mathrm{d}V$ as discussed in section 5.4. Under isothermal conditions, the work extracted $\gtreqless \Delta U$ depending on whether heat is absorbed or rejected in the change. Thus F becomes a useful energy function for isothermal processes. In an isovolumic change, the change in F is related to the change in temperature.

Gibbs function. The importance of the Gibbs potential is that it remains constant in reversible processes occurring under isothermal, isobaric conditions. These are the conditions applying to many physical and chemical changes. The constancy of the Gibbs function may then be used to represent the system constraints. We shall later develop its application to the determination of the equilibrium states of systems containing several phases (chapter 10) and several components (chapter 11).

As may be seen from their definitions, all the thermodynamic potential functions are extensive quantities.

7.2. The Legendre Differential Transformation

In systems with more than two degrees of freedom there are correspondingly more thermodynamic potentials and their differentials contain correspondingly more terms. As in the case of the two-parameter system, one first constructs the expression for the differential of the internal energy and from that derives the other potentials.

For a system with n degrees of freedom, the expression for $\mathrm{d}U$ contains $T\,\mathrm{d}S$ and $(n-1)$ work-like terms, each of the form $X_i\,\mathrm{d}x_i$. The system therefore has $2n$ *primary variables* forming n conjugate pairs whose products have the dimensions of energy. (Pairs like T, S; p, V; $\mathbf{E}$, $\mathbf{p}$.) The potential functions for the system with two degrees of freedom, which were discussed in the last section, correspond to all the possible combinations of independent variables when one is taken from each conjugate pair. Thus, for a system with n degrees of freedom, there will be 2^n potential functions corresponding to the twofold choice offered by each pair. For example, a wire under tension and for which volume changes are important, has three pairs of primary variables: T, S; p, V; $\mathscr{F}, L$. The eight potential functions will correspond to the following sets of independent variables.

$$\begin{array}{llll} T, p, \mathscr{F} & T, V, \mathscr{F} & S, p, \mathscr{F} & S, V, \mathscr{F} \\ T, p, L & T, V, L & S, p, L & S, V, L. \end{array}$$

It is clearly a great advantage to have a systematic way of generating these potentials as and when required. The simplest method is the following.

Firstly, the expression for $\mathrm{d}U$ is written down. This consists of $T\,\mathrm{d}S$ plus all the work terms and has as its independent variables the extensive members of the conjugate pairs. To obtain a potential with a different set of independent variables, one picks out the terms in which the wrong member of the pair is the independent variable and adds to or subtracts from $\mathrm{d}U$ the differential of the *product* of the conjugate pair so as to remove the unwanted term and replace it by the required one. This produces a new differential expression, still with n terms but, with a different set of independent variables. Obviously, it is exact, because it was obtained by adding together exact differentials, namely, $\mathrm{d}U$ and terms like $\mathrm{d}(pV)$. It also has the dimensions of energy and is therefore the differential of a new potential function. This procedure is known as a *Legendre differential transformation.* We illustrate it by returning to the example of the wire subject to work by tension and hydrostatic pressure.

For the wire, the first law becomes

$$\mathrm{d}U = T\,\mathrm{d}S + \mathscr{F}\,\mathrm{d}L - p\,\mathrm{d}V$$

which has as independent variables S, L, and V. Suppose we wish to construct the potential with proper variables T, L, and p. Then the first and last terms need to be transformed. We may effect this by adding $-\mathrm{d}(TS) + \mathrm{d}(pV)$. This generates the differential of the new potential:

$$\begin{aligned} \mathrm{d}G'(T, L, p) &= \mathrm{d}U - \mathrm{d}(TS) + \mathrm{d}(pV) \\ &= -S\,\mathrm{d}T + \mathscr{F}\,\mathrm{d}L + V\,\mathrm{d}p \end{aligned}$$

from which we see that the new potential is

$$G' = U - TS + pV.$$

Although, in section 7.1, we *stated* the four thermodynamic potentials for the system subject to work by hydrostatic pressure only, we could have generated them from the expression for dU by applying Legendre differential transformations to obtain all the possible combinations of independent variables.

7.3. The Maxwell Relations

For systems with two degrees of freedom, there are four extremely useful equations which relate partial differentials of the fundamental thermodynamic variables. They may be deduced from the differential forms of the thermodynamic potentials. Their usefulness lies in the transformation of variables they make possible.

If we form partial differentials of U with respect to its proper variables we obtain

$$\left(\frac{\partial U}{\partial S}\right)_V = T \quad \text{and} \quad \left(\frac{\partial U}{\partial V}\right)_S = -p.$$

Differentiating again with respect to the opposite variables:

$$\frac{\partial^2 U}{\partial V \partial S} = \left(\frac{\partial T}{\partial V}\right)_S \quad \text{and} \quad \frac{\partial^2 U}{\partial S \partial V} = -\left(\frac{\partial p}{\partial S}\right)_V.$$

But

$$\frac{\partial^2 U}{\partial V \partial S} = \frac{\partial^2 U}{\partial S \partial V}, \qquad \text{(cf. equation (1.9))}$$

so that

$$\left(\frac{\partial T}{\partial V}\right)_S = -\left(\frac{\partial p}{\partial S}\right)_V.$$

The same result may also be obtained immediately by using the condition for dU to be an exact differential. (Equation (1.11).) The differential form of U is

$$dU = T\,dS - p\,dV.$$

Applying the condition to the coefficients on the right-hand side, we obtain immediately

$$\left(\frac{\partial T}{\partial V}\right)_S = -\left(\frac{\partial p}{\partial S}\right)_V.$$

Following the same procedure with H, F, and G we obtain three more equations of a similar form. The four equations are known as the *Maxwell Relations*. They are (from U, G, H, and F respectively)

$$\left.\begin{aligned} &\left(\frac{\partial T}{\partial V}\right)_S = -\left(\frac{\partial p}{\partial S}\right)_V \\ &\left(\frac{\partial T}{\partial V}\right)_p = -\left(\frac{\partial p}{\partial S}\right)_T \\ &\left(\frac{\partial T}{\partial p}\right)_S = \left(\frac{\partial V}{\partial S}\right)_p \\ &\left(\frac{\partial T}{\partial p}\right)_V = \left(\frac{\partial V}{\partial S}\right)_T . \end{aligned}\right\} \quad (7.3)$$

These may be recalled easily by remembering the following rules:

1. Cross multiplication of the variables always gives the form $(TS)=(pV)$ with the dimensions of energy.
2. Opposite pairs of variables are constant.
3. The sign is positive if T appears with p (for positive).

If we chose to introduce differential coefficients of the potential functions, we could derive many other equalities between differential coefficients like

$$\left(\frac{\partial U}{\partial S}\right)_V = \left(\frac{\partial H}{\partial S}\right)_p ;$$

but unlike the Maxwell relations, these are rarely useful and it is easier to deduce them if they are needed.

Once again, although we have deduced the Maxwell relations in the form appropriate to a system subjected to work by hydrostatic pressure, similar equations hold for any two-parameter system. S and T, of course, apply to any system, but p and V have to be replaced by their corresponding variables:

$$-p,\ V \rightarrow \mathscr{F}, L;\ \gamma, A;\ B, m;\ \text{etc.}$$

(See Table 3.1.)

In systems with more than two independent variables, the number of Maxwell relations becomes much greater. The system with n independent variables has 2^n potential functions (section 7.2), and each of these yields $n(n-1)/2$ Maxwell relations. There are systematic methods for deriving them[2]; but for most purposes it is easier to consider each problem individually, constructing when needed the potential functions which yield the required differential coefficients. We shall later consider two simple cases with $n=3$. (Sections 8.6 and 8.7.)

[2] See F. H. Crawford, *Heat, Thermodynamics and Statistical Physics*, pp. 370–2, Harcourt, Brace and World, 1963.

8. Applications to Simple Systems

In this chapter, we shall primarily be concerned with the application of thermodynamics to simple systems undergoing reversible changes. The topics have been chosen to illustrate the general methods of thermodynamics as well as for their intrinsic importance.

8.1. Some Properties of Specific Heats

We have shown that for reversible changes it is possible to express $đQ$ in terms of state variables: $đQ_{\text{rev}} = T\,\mathrm{d}S$. This renders $đQ_{\text{rev}}$ amenable to manipulation by the techniques applicable to functions of state. In particular, thermodynamic coefficients involving $đQ_{\text{rev}}$ may be expressed in terms of complete differentials which makes transformation of variables a straightforward matter. We may illustrate this by deriving some important general results for the principal specific heats of a system subject to work by hydrostatic pressure only.

The principal specific heats are the heat capacities for change of temperature where the constraints correspond to constancy of primary variables. For a system subject to one kind of work only, there are two. In the case of work by hydrostatic pressure they are c_p and c_v. These are related through the definition of heat capacities (section 3.6) to the entropy as follows:

$$\left.\begin{aligned} c_p &= \frac{đQ_p}{\mathrm{d}T} = T\left(\frac{\partial s}{\partial T}\right)_p \\ c_v &= \frac{đQ_v}{\mathrm{d}T} = T\left(\frac{\partial s}{\partial T}\right)_v. \end{aligned}\right\} \quad (8.1)$$

Certain differentials of the principal specific heats may be expressed in simple forms involving directly observable functions of state:

$$\begin{aligned} \left(\frac{\partial c_p}{\partial p}\right)_T &= \left.\frac{\partial}{\partial p}\right)_T \left\{T\left(\frac{\partial s}{\partial T}\right)_p\right\} \\ &= T\left.\frac{\partial}{\partial p}\right)_T \left(\frac{\partial s}{\partial T}\right)_p, \end{aligned}$$

since T remains constant under the partial differentiation wit respect to p. Reversing the order of differentiation,

$$=T \left.\frac{\partial}{\partial T}\right)_p \left(\frac{\partial s}{\partial p}\right)_T,$$

and using a Maxwell relation,

$$= -T \left.\frac{\partial}{\partial T}\right)_p \left(\frac{\partial v}{\partial T}\right)_p$$

$$= -T \left(\frac{\partial^2 v}{\partial T^2}\right)_p.$$

Thus

$$\left(\frac{\partial c_p}{\partial p}\right)_T = -T \left(\frac{\partial^2 v}{\partial T^2}\right)_p. \quad (8.2$$

A similar result follows in the same way for the differential of c with respect to v:

$$\left(\frac{\partial c_v}{\partial v}\right)_T = +T \left(\frac{\partial^2 p}{\partial T^2}\right)_v. \quad (8.3$$

In the case of the perfect gas (section 8.2) for which $pV = RT$, both these coefficients are zero.

It is also possible to obtain general forms for the difference of the principal specific heats. The most useful of these is that which expresses the difference in terms of the expansion coefficient and the compressibility. This is obtained directly by expanding s as a function of T and v or of T and p.

$$s = s(T, v)$$

$$\mathrm{d}s = \left(\frac{\partial s}{\partial T}\right)_v \mathrm{d}T + \left(\frac{\partial s}{\partial v}\right)_T \mathrm{d}v.$$

Whence

$$\left(\frac{\partial s}{\partial T}\right)_p = \left(\frac{\partial s}{\partial T}\right)_v + \left(\frac{\partial s}{\partial v}\right)_T \left(\frac{\partial v}{\partial T}\right)_p.$$

Substituting for the temperature derivatives of the entropy with equations (8.1),

$$c_p - c_v = T \left(\frac{\partial s}{\partial v}\right)_T \left(\frac{\partial v}{\partial T}\right)_p$$

and applying a Maxwell relation,

$$c_p - c_v = T\left(\frac{\partial p}{\partial T}\right)_v \left(\frac{\partial v}{\partial T}\right)_p . \tag{8.4}$$

The expansivity and compressibility are defined as:

isobaric cubic expansivity, $\beta_p = \frac{1}{V}\left(\frac{\partial V}{\partial T}\right)_p$

isothermal compressibility, $\kappa_T = -\frac{1}{V}\left(\frac{\partial V}{\partial p}\right)_T .$

Substituting in (8.4) and using

$$\left(\frac{\partial p}{\partial T}\right)_v = -\left(\frac{\partial p}{\partial v}\right)_T \left(\frac{\partial v}{\partial T}\right)_p ,$$

we obtain

▶ $$c_p - c_v = vT\,\frac{\beta^2}{\kappa_T} . \tag{8.5}$$

If (8.4) is applied to 1 mol of a perfect gas, we obtain

▶ $$C_{mp} - C_{mV} = R . \tag{8.6}$$

8.1.1. The Isothermal–Adiabatic Transformation of Moduli

The differential of an intensive variable with respect to its associated extensive variable is called a *stiffness* coefficient. The reciprocal differential is a *compliance* coefficient. These are important physical quantities and like all thermodynamic coefficients they are partial differentials since their values depend on the conditions under which they are measured. Two common constraints are that the system is kept isothermal or thermally isolated. Thus we have already used the isothermal compressibility and we may similarly define an adiabatic compressibility,

$$\kappa_S = -\frac{1}{V}\left(\frac{\partial V}{\partial p}\right)_S$$

where we have assumed that the changes are thermodynamically reversible and have replaced the condition of thermal isolation by constancy of entropy. A very simple relationship exists between the ratio of the isothermal and adiabatic coefficients and the principal specific heats. We show this for the case of the compressibilities.

$$\frac{\kappa_T}{\kappa_S}=\frac{\left(\frac{\partial V}{\partial p}\right)_T}{\left(\frac{\partial V}{\partial p}\right)_S}$$

$$=\frac{\left(\frac{\partial V}{\partial T}\right)_p\left(\frac{\partial T}{\partial p}\right)_V}{\left(\frac{\partial V}{\partial S}\right)_p\left(\frac{\partial S}{\partial p}\right)_V} \qquad \text{(reciprocity theorem)}$$

$$=\frac{\left(\frac{\partial S}{\partial V}\right)_p\left(\frac{\partial V}{\partial T}\right)_p}{\left(\frac{\partial S}{\partial p}\right)_V\left(\frac{\partial p}{\partial T}\right)_V} \qquad \text{(rearranging)}$$

$$=\frac{\left(\frac{\partial S}{\partial T}\right)_p}{\left(\frac{\partial S}{\partial T}\right)_V}$$

and substituting with the specific heats,

▶ $$\frac{\kappa_T}{\kappa_S}=\frac{c_p}{c_v}=\gamma. \qquad (8.7)$$

Similar results hold for coefficients formed from other pairs of variables. For example,

$$\left.\begin{array}{lll}\text{permittivities,} & \varepsilon=\left(\frac{\partial D}{\partial E}\right): & \frac{\varepsilon_T}{\varepsilon_S}=\frac{c_E}{c_D} \\ \text{Young moduli,} & E=\frac{L}{A}\left(\frac{\partial \mathscr{F}}{\partial L}\right): & \frac{E_T}{E_S}=\frac{c_L}{c_{\mathscr{F}}} \\ \text{magnetic susceptibilities,} & \chi=\left(\frac{\partial M}{\partial H}\right): & \frac{\chi_T}{\chi_S}=\frac{c_H}{c_M}.\end{array}\right\} \qquad (8.8)$$

8.2. The Perfect Gas

8.2.1. Definition of the Perfect Gas

The perfect gas is most conveniently defined as one which obeys both Boyle's law and Joule's law. These state:

Boyle's Law: If the temperature is maintained constant then the product pV is constant.

Joule's Law: The internal energy is independent of p and V.

These laws are not obeyed strictly by any real gas. In the case of Boyle's law, accurate isothermal measurements of the product pV may be fitted well by a power series in p:[1]

$$pV = A + Bp + Cp^2 + Dp^3 + \dots.$$

(Such an expression is known as a *virial expansion* and the coefficients as *virial coefficients*.) The coefficients are functions of temperature and may be of either sign. Clearly, by going to the limit of small p, a real gas may be made to obey Boyle's law to any desired accuracy. Away from this limit, Boyle's law is best obeyed if the second virial coefficient vanishes leaving only second and higher order corrections to pV. The temperature at which this occurs is known as the *Boyle temperature*.

In the original experiments on which Joule's law is based, gases were made to perform a 'free expansion', that is, one in which no external work is performed and no heat enters or leaves the system. (See section 9.1.) To the accuracy of the experiments, no temperature change occurred. The constraint of the experiment was

$$\mathrm{d}U = đQ + đW = 0,$$

and the result that

$$\left(\frac{\partial T}{\partial V}\right)_U = -\left(\frac{\partial T}{\partial U}\right)_V \left(\frac{\partial U}{\partial V}\right)_T = 0.$$

But

$$\left(\frac{\partial U}{\partial T}\right)_V = C_V$$

which is a well behaved finite quantity. Thus it follows from Joule's experiments that

$$\left(\frac{\partial U}{\partial V}\right)_T = 0.$$

Similarly,

$$\left(\frac{\partial U}{\partial p}\right)_T = \left(\frac{\partial U}{\partial V}\right)_T \left(\frac{\partial V}{\partial p}\right)_T.$$

Again, $(\partial V/\partial p)_T$ is a well behaved finite quantity (which may be calculated from the equation of state), so that we also have

$$\left(\frac{\partial U}{\partial p}\right)_T = 0.$$

[1] Onnes, *Communs. Kamerlingh Onnes Lab., Leiden* No. 71, 1901. Values for the virial coefficients may be found in the *Handbuch der Experimentalphysik*, edited by S. Flügge, **VIII**, Part 2, Springer-Verlag, 1963.

Thus U must be a function of temperature only.

Again, Joule's law is only an approximation to the low pressure behaviour of real gases. More accurate experiments[2] show that the behaviour of U is better described by adding a term linear in p:

$$U(p, T) = U(0, T) + \alpha p.$$

Thus $(\partial U/\partial p)_T$ is, in fact, roughly constant even in the low pressure limit. The reason why Joule's law is nevertheless a reasonable description of the limiting behaviour of a real gas is that by making the pressure small, the part of the internal energy which varies with volume or pressure may be made as small a *proportion* of the total as desired. The reasons for the departures from Boyle's and Joule's laws will be discussed in section 8.3.

8.2.2. The Equation of State of the Perfect Gas

From Boyle's and Joule's laws we may immediately derive the equation of state for a perfect gas. From the first law,

$$\begin{aligned}\left(\frac{\partial U}{\partial V}\right)_T &= T\left(\frac{\partial S}{\partial V}\right)_T - p \\ &= T\left(\frac{\partial p}{\partial T}\right)_V - p \\ &= 0 \qquad \text{by Joule's law.}\end{aligned}$$

Thus

$$\left(\frac{\partial p}{\partial T}\right)_V = \frac{p}{T},$$

and integrating,

$$\ln p = \ln T + f(V),$$

where $f(V)$ is an arbitrary function of V which appears in place of an integration constant since V is held constant in the differentiation.[3] Rearranging this result

$$pF(V) = T. \tag{8.9}$$

But we know from Boyle's law that if the temperature is constant

[2] See, for example, F. D. Rossini and M. Fransden, *J. Res. Nat. Bur. Stand.* **9**, 733, 1932.

[3] Integration of a partial differential equation always introduces arbitrary functions of the variables held constant in the differentiation. As in this case, they can often be determined by reference to other known properties of the system.

then pV is constant so that $F(V) \propto V$. Thus for the perfect gas the equation of state must be of the form

$$pV_m = RT \tag{8.10}$$

where R is called the *gas constant.*[4] Its value is

$$R = 8{\cdot}32 \text{ J K}^{-1} \text{ mol}^{-1}.$$

In the limit of low pressure, real gases approach the ideal behaviour for reasons which we shall discuss in section 8.3. Thus, temperatures determined by gas thermometry in the low pressure limit are thermodynamic temperatures provided that the size of the unit is chosen correctly. (Section 2.5.) This is why the determination of thermodynamic temperatures is usually based on gas thermometry.

8.2.3. The Adiabatic Equation of the Perfect Gas

The adiabatic equation of the perfect gas may be derived directly from Boyle's law using the general result for isothermal–adiabatic transformation of moduli which was derived in section 8.1.1. We showed that for reversible adiabatic changes

$$\frac{\left(\dfrac{\partial V}{\partial p}\right)_T}{\left(\dfrac{\partial V}{\partial p}\right)_S} = \frac{c_p}{c_V} = \gamma. \qquad \text{(equation (8.7))}$$

Boyle's law states that for constant temperature, pV is constant.

[4] It should be noted that (8.9) is essentially Charles' law which states that if the volume of a gas is kept constant, the pressure is proportional to the temperature. One might think that it would have been acceptable to choose Charles' law instead of Joule's law in setting up our definition of the perfect gas; but this would have been less satisfactory: one would then have had to use the perfect gas in a Carnot cycle to demonstrate that the temperature scale so defined was the same as that we had arrived at in the discussion of the second law. Boyle's and Joule's laws involve *constancy* of temperature and therefore do not involve us in particular scales. The temperature which eventually emerges in the equation of state comes out of the thermodynamics with which we treat Joule's law and is therefore necessarily thermodynamic temperature. The temperature *implied* in Charles' law is the perfect gas absolute scale. (Section 2.4.) We should thus still have been left to demonstrate that the temperature appearing in the equation of state was also thermodynamic temperature.

Taking logs and differentiating,

$$\frac{\mathrm{d}p}{p}+\frac{\mathrm{d}V}{V}=0$$

or,

$$\left(\frac{\partial p}{\partial V}\right)_T=-\frac{p}{V}.$$

Substituting in the transformation,

$$\left(\frac{\partial p}{\partial V}\right)_S=-\gamma\frac{p}{V}.$$

If the heat capacities are constant, this may be integrated to give

$$\ln p+\gamma \ln V=\text{const., or}$$

$$\blacktriangleright \qquad pV^\gamma=\text{const.} \tag{8.11}$$

8.2.4. The Entropy of the Perfect Gas

The equation of state of the perfect gas may be used to express the entropy in terms of any two of the fundamental variables p, V, T. For example, expanding S in terms of p and T,

$$S=S(p,T)$$

$$\mathrm{d}S=\left(\frac{\partial S}{\partial p}\right)_T\mathrm{d}p+\left(\frac{\partial S}{\partial T}\right)_p\mathrm{d}T$$

$$=-\left(\frac{\partial V}{\partial T}\right)_p\mathrm{d}p+\left(\frac{\partial S}{\partial T}\right)_p\mathrm{d}T$$

where we have transformed the first term with a Maxwell relation. So far this equation applies to any system subject to work by hydrostatic pressure. We now use the equation of state. From (8.10),

$$p\left(\frac{\partial V_m}{\partial T}\right)_p=R.$$

Also,

$$T\left(\frac{\partial S}{\partial T}\right)_p=C_p$$

so that for one mole we have

$$\mathrm{d}S_m=-\frac{R}{p}\mathrm{d}p+\frac{C_{mp}}{T}\mathrm{d}T.$$

Integrating this complete differential expression, and taking C_{mp} to be independent of temperature,

$$S_m(p,T)=S_0-R\ln p+C_{mp}\ln T. \tag{8.12}$$

S may be expressed in terms of p and T or of p and V either by direct expansion of dS in terms of these variables, or by substituting in (8.12) using the equation of state and the expression for the difference of the principal heat capacities. The results are:

$$S_m(V, T) = S_0' + R \ln V + C_{mV} \ln T \qquad (8.13)$$

$$S_m(p, V) = S_0'' + C_{mV} \ln p + C_{mp} \ln V. \qquad (8.14)$$

Equations (8.12) and (8.13) appear to predict that $S \to -\infty$ as $T \to 0$. The explanation of this seemingly absurd result is simply that the perfect gas laws cannot be obeyed by any real substance to absolute zero. In chapter 12, we shall see that the third law requires that all specific heats vanish as $T \to 0$. Microscopically, this is a consequence of the quantization of energy, for as kT becomes comparable with the spacing of the lowest energy levels, equipartition of energy will break down and eventually the specific heat will vanish exponentially.[5] This ensures that the entropy remains finite and, by definition, positive. For further discussion, see chapter 12.

8.3. The Behaviour of Real Pure Substances

The equation of state of the perfect gas may be deduced from the postulates of the kinetic theory of gases. The essential microscopic assumptions behind the perfect gas equation are that the gas molecules occupy a negligible volume (i.e., they approximate to point masses), and that there are no intermolecular forces. Real gases differ from the perfect gas because neither of these assumptions is true, although in the limit of low pressure, when the gas molecules are far apart, the forces between them and the volume they occupy become negligible and the gas does approach the ideal behaviour.

Various modifications to the perfect gas laws have been suggested as providing a better description of real gases. The best known of these are[6]

Van der Waals' equation: $$\left(p + \frac{a}{V_m^2}\right)(V_m - b) = RT$$

[5] This is true even for a Fermi gas. The linear specific heat which holds for electrons in a metal down to the lowest temperatures which can be reached is the result of the *extremely* close spacing of the energy levels. For a small sample, their separation is about 10^{-16} eV, corresponding to a temperature of about 10^{-12} K.

[6] For further discussion of these, see, J. Jeans, *Kinetic Theory of Gases*, chapter 3, Cambridge University Press, 1962.

Dieterici's equation: $p(V_m-b')=RT\exp\left[-\frac{a'}{RTV_m}\right].$

Both may be derived by kinetic arguments using different approximations. The constants b and b' take account of the finite size of the molecules by decreasing the effective available volume. The constants a and a' are associated with the attractive forces between the molecules.

The attractive forces between the molecules of a real gas may be thought of as constituting an 'internal pressure' against which 'internal work' is done during expansion. This may be seen by considering the heat capacity for change of volume,

$$C_T^{(V)}=\frac{đQ_T}{dV},$$

being the heat absorbed per unit increase in volume when the system is kept at a constant temperature. Using the first law, this becomes

$$C_T^{(V)}=p+\left(\frac{\partial U}{\partial V}\right)_T.$$

In terms of the kinetic theory, the kinetic energies of the molecules must remain constant in such an expansion. The heat absorbed must therefore be required for other forms of work. The first term is simply the external work done, $đW_{\text{ext}}=p\,dV$, while the second must be internal to the system, and can only be associated with the process of increasing the separation of the molecules. That is, it corresponds to work done against the intermolecular forces. This term is absent only if Joule's law is satisfied, so that we see that the assumption that Joule's law is true, is equivalent to the microscopic assumption that the intermolecular forces are negligible.

In detail, it is not possible entirely to separate the concepts of intermolecular forces and finite molecular size. However, if both these were absent, there would be no condensed state of matter, and the perfect gas laws would be true at all temperatures and pressures. The p–V–T relation for a real pure substance is represented in Fig. 8.1. At high temperatures, the isotherms approximate towards the rectangular hyperbolae of Boyle's law. (For example, AE.) As the temperature is reduced they become more distorted until, at the *critical temperature*, a horizontal portion appears corresponding to the appearance of a distinguishable liquid phase. The point C is

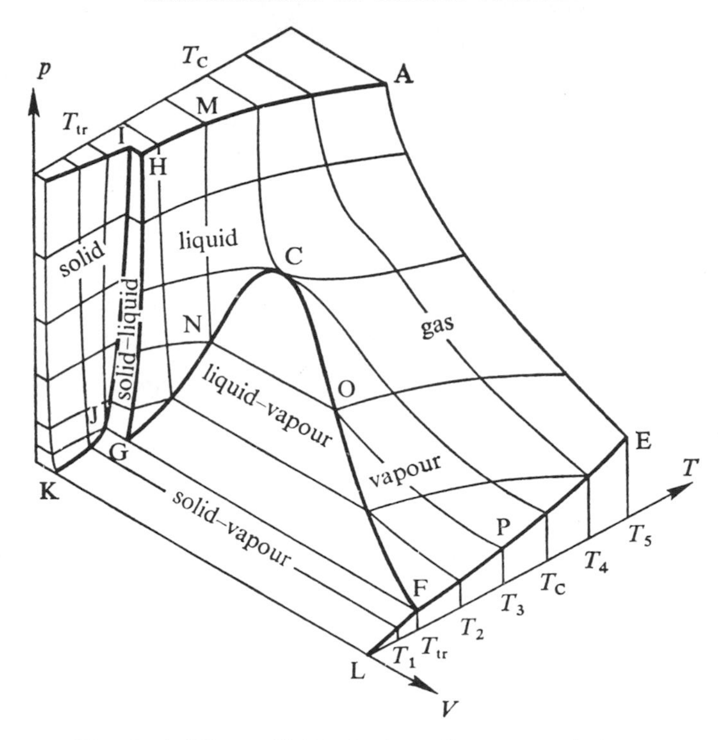

Fig. 8 . 1. The p–V–T relation of a pure substance.

known as the *critical point*. It corresponds to the conditions

$$\left(\frac{\partial p}{\partial V}\right)_T = \left(\frac{\partial^2 p}{\partial V^2}\right)_T = 0$$

and for each substance is associated with unique values of the temperature, pressure, and volume. These are known as the *critical constants*. They are listed for several gases in Table 8.1.

Below the critical temperature, liquid and vapour[7] may coexist in the region GCF of Fig. 8.1. Here the isotherms are horizontal (the system is infinitely compressible) and passage along the isotherm in the direction of decreasing V corresponds to liquefaction of the vapour. Thus, along PO only vapour exists. At O the liquid phase appears, and along ON further compression causes the vapour to

[7] It is convenient to use the term *vapour* to distinguish the region of the gaseous phase which lies below the critical temperature. Thus, a vapour always undergoes a change of phase on (isothermal) compression.

liquefy until, at N, all the substance has been converted to the liquid and further compression gives a rapid rise in pressure along NM corresponding to the greater modulus of the liquid. Further decrease of volume eventually results in a transition to the solid phase via the continuation of the region GHIJ in which solid and liquid coexist.[8] At still lower temperatures, such as T_1, the vapour on compression passes into the solid phase and in the region KJGFL solid and vapour coexist. Only at one temperature, the *triple temperature* T_{tr}, may all three phases be present together.

Table 8.1. The Critical Constants for Some Gases*

gas	p_c/atm	$V_c/10^{-6}$ m^3 mol^{-1}	T_c/K
helium	2·25	61·55	5·2
hydrogen	12·8	69·68	33·2
nitrogen	33·49	90·03	125·97
argon	47·996	77·07	150·66
carbon dioxide	72·83	94·23	304·16

* From E. A. Moelwyn-Hughes, *Physical Chemistry*, p. 573, Pergamon Press, 1957.

Three-dimensional representations of the equation of state such as is illustrated in Fig. 8.1 are complicated and contain much more information than is usually needed. It is often sufficient to project sections of the p–V–T surface on to one of the principal planes. This is illustrated in Fig. 8.2 for the two most useful projections. Since, for a given temperature, liquefaction or solidification proceed at constant pressure, the mixed phase regions project into lines on the p–T plane. On the p–V plane, the mixed phase states are distinguishable, giving a region of horizontal isotherms.

Clearly, the p–V–T relationship for a pure substance, like that illustrated in Fig. 8.1, cannot be represented algebraically. The best one can hope to do is to construct approximate equations, like the van der Waals equation for the gas phase, which take some account of the deviations from simple behaviour near the mixed phase states. In chapter 10, we shall return to a more detailed discussion of change of phase.

[8] In water, where the liquid expands on solidification the solid–liquid transition slopes towards the p-axis in the direction of increasing p so that isothermal compression of the liquid can never lead to solidification. Instead, there exist temperatures for which compression leads from vapour to solid and then to liquid.

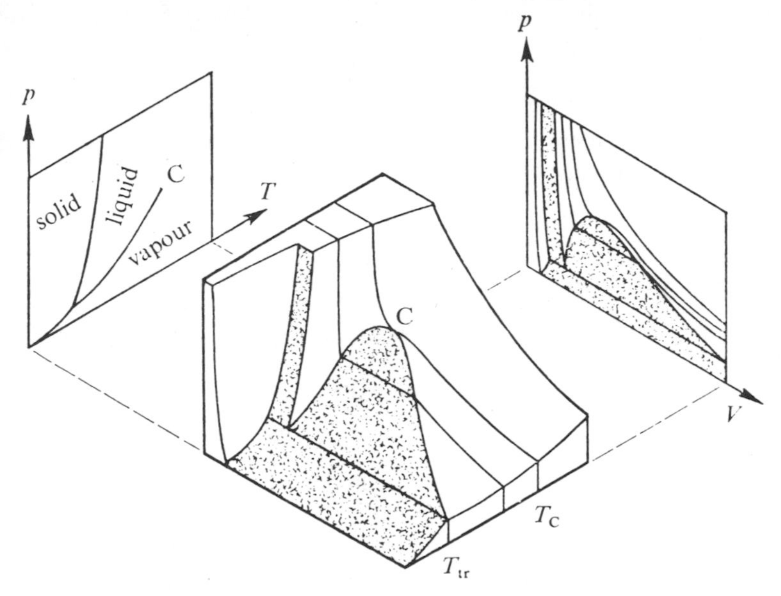

Fig. 8.2. The two most useful projections of the p–V–T surface.

8.4. The Elastic Rod or Filament

Usually, when a wire or rod is stretched the only significant work term is that associated with the tension. In particular, work by hydrostatic pressure is usually negligible. Under these conditions the system has two degrees of freedom and the first law takes the form

$$\mathrm{d}U = T\,\mathrm{d}S + \mathscr{F}\,\mathrm{d}L. \tag{8.15}$$

8.4.1. Heating at Constant Length

Clearly, a thin wire which is initially at zero tension cannot be heated at constant length, for it would simply expand and buckle. If, however, it is initially under tension and is attached to rigid supports it may be heated at constant length and a change in tension will result. Similarly, if a rigid rod is constrained between rigid supports a rise in temperature will result in a change in compressional force. This is the significance attached to the coefficient $(\partial\mathscr{F}/\partial T)_L$. This quantity may be expressed in terms of well known variables.

$$\left(\frac{\partial\mathscr{F}}{\partial T}\right)_L = -\left(\frac{\partial\mathscr{F}}{\partial L}\right)_T\left(\frac{\partial L}{\partial T}\right)_{\mathscr{F}}. \tag{8.16}$$

By their definitions, the isothermal Young modulus E_T is given by

$$\left(\frac{\partial \mathscr{F}}{\partial L}\right)_T = \frac{AE_T}{L}$$

where A is the cross-sectional area of the wire or rod, and the linear expansivity at constant tension, $\alpha_{\mathscr{F}}$, is given by

$$\alpha_{\mathscr{F}} = \frac{1}{L}\left(\frac{\partial L}{\partial T}\right)_{\mathscr{F}}.$$

Then (8.16) becomes

$$\left(\frac{\partial \mathscr{F}}{\partial T}\right)_L = -AE_T\alpha_{\mathscr{F}}. \tag{8.17}$$

For a small change in temperature, the change in tension is

$$\Delta\mathscr{F} = -AE_T\alpha_{\mathscr{F}}\Delta T.$$

Note that this does not depend on the length of the rod.

8.4.2. The Heat Absorbed on Isothermal Extension

If the rod or filament is extended isothermally and reversibly, the heat absorbed per unit extension is

$$C_T^{(L)} = \frac{đQ_T}{\mathrm{d}L} = T\left(\frac{\partial S}{\partial L}\right)_T = -T\left(\frac{\partial \mathscr{F}}{\partial T}\right)_L$$

where we have used the Maxwell relation which comes from the differential of the Helmholtz function, $F = U - TS$. With (8.17), this becomes

$$C_T^{(L)} = AE_TT\alpha_{\mathscr{F}}. \tag{8.18}$$

A, E_T, and T are, of course, necessarily positive, and most materials expand on heating so that α is normally positive. Thus, heat is usually absorbed on isothermal extension. An exception is rubber. We have already discussed the microscopic explanation of this difference in section 5.6.4.

If the system is deformed adiabatically, the resulting temperature change is simply related to the quantity we have just derived. Provided that the deformation is reversible we have

$$\left(\frac{\partial T}{\partial L}\right)_S = -\left(\frac{\partial T}{\partial S}\right)_L\left(\frac{\partial S}{\partial L}\right)_T = -\frac{T\left(\frac{\partial S}{\partial L}\right)_T}{T\left(\frac{\partial S}{\partial T}\right)_L} = -\frac{AE_TT\alpha_{\mathscr{F}}}{C_L} \tag{8.19}$$

where C_L is the heat capacity at constant length.

8.5. The Reversible Electric Cell

Dry cells and the lead–acid accumulator are far from reversible but there are many cells based on simple chemical reactions which are very nearly reversible. Typical of these is the Daniel cell which is illustrated in Fig. 8.3. The electrodes are of copper and zinc and each is immersed in a saturated solution of its sulphate. The inter-diffusion of the two sulphate solutions is inhibited by the porous partition which divides the cell. The copper electrode is found to be positive with respect to the zinc so that discharging involves the passage of conventional current from copper to zinc (i.e., electrons flow from the zinc to the copper). In solution, both zinc and copper are doubly ionized, so that discharge involves the passage of zinc into solution to form Zn^{++} ions and the discharge of Cu^{++} ions at the copper electrode to form neutral copper which is deposited. On charging the reverse process takes place. The chemical reaction is

$$Cu + Zn^{++} \rightleftharpoons Cu^{++} + Zn.$$

For such cells to be reversible, the process of charging or discharging must be carried out sufficiently slowly for Joule heating to be negligible.[9] Also, for the chemical reaction to be reversible, there must normally be no gas evolved. In this case, volume changes will be negligible and the cell will be subject to work by charging only. Generally, the e.m.f. will be a function of the temperature and the charge Z: $\mathscr{E} = \mathscr{E}(T, Z)$; but if the solutions are kept saturated, the e.m.f. becomes independent of the extent of charging and is a function of temperature only.

For such cells, the first law becomes

$$dU = T\,dS + \mathscr{E}\,dZ \tag{8.20}$$

where Z is measured in the positive direction during charging. If we express dS in terms of dT and dZ this becomes

$$dU = T\left(\frac{\partial S}{\partial T}\right)_Z dT + \left[\mathscr{E} + T\left(\frac{\partial S}{\partial Z}\right)_T\right] dZ. \tag{8.21}$$

[9] The *rate* of Joule heating is i^2R where i is the current and R the resistance of the cell. Hence for the passage of a given charge Z in time τ the total joule heating is Z^2R/τ which may be made as small as one wishes by performing the charging slowly and making τ large. The electrical work done on the cell, $\int \mathscr{E}\,dZ$, is unaffected by the rate at which the charging is performed.

The two entropy terms are heat capacities. The first, $T(\partial S/\partial T)_Z$, is the heat capacity at constant charge $C_Z^{(T)}$; and the second is the heat of charging:

$$T\left(\frac{\partial S}{\partial Z}\right)_T = \frac{đQ_T}{\mathrm{d}Z} = C_T^{(Z)}.$$

It is the rate at which heat is absorbed during a reversible isothermal increase in charge. This term may be transformed to a more useful form by using a Maxwell relation. We may obtain this either by substitution of the appropriate variables in the standard form (see equations (7.3)); or by deriving it from the appropriate potential

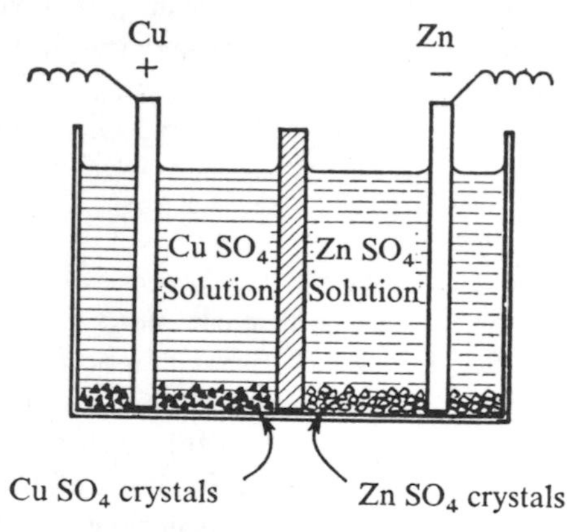

Fig. 8.3. The Daniel cell. The crystalline salts keep the solutions saturated. The porous partition inhibits the interdiffusion of the two solutions.

function. To derive it, we must construct the differential of the potential function with Z and T as proper variables. (These are the independent variables in the coefficient we wish to transform.) We obtain this from $\mathrm{d}U$ by the Legendre transformation (section 7.2)

$$\mathrm{d}F = \mathrm{d}U - \mathrm{d}(TS) = -S\,\mathrm{d}T + \mathscr{E}\,\mathrm{d}Z.$$

Following the methods we used in section 7.3 to derive the fundamental Maxwell relations, we now either differentiate in both orders with respect to T and Z and equate, or simply apply the condition for $\mathrm{d}F$ to be exact. Either gives immediately

$$\left(\frac{\partial S}{\partial Z}\right)_T = -\left(\frac{\partial \mathscr{E}}{\partial T}\right)_Z. \qquad (8.22)$$

Substituting in (8.21) we obtain

$$\mathrm{d}U = T\left(\frac{\partial S}{\partial T}\right)_Z \mathrm{d}T + \left[\mathscr{E} - T\left(\frac{\partial \mathscr{E}}{\partial T}\right)_Z\right] \mathrm{d}Z.$$

For the cell in which the solutions are kept saturated we remember that $\mathscr{E}$ is a function of T alone and this becomes

$$\mathrm{d}U = T\left(\frac{\partial S}{\partial T}\right)_Z \mathrm{d}T + \left[\mathscr{E} - T\frac{\mathrm{d}\mathscr{E}}{\mathrm{d}T}\right] \mathrm{d}Z. \quad (8.23)$$

Now most chemical reactions take place at constant pressure so that it is convenient to measure energy changes in terms of the enthalpy, $H = U + pV$, for which the proper variables are S, Z, and p. When reaction occurs the total enthalpy change (the enthalpy of the products less that of the reactants consumed) is called the *heat of reaction*. In a reaction which does not involve electrical work, the enthalpy change is entirely due to exchange of heat with the surroundings (hence the misleading name *heat content* for H); but for the electric cell, it results partly from exchange of heat and partly from electrical work.

For our simple cells in which there is no volume change, we have, for isobaric conditions, $\Delta H = \Delta U$. The total charge required to cause 1 gram-atom of one of the metals to pass into solution is $\Delta Z = zF'$ where z is the valence of the metal ion and $F' = Ne$ is the faraday ($9{\cdot}65 \times 10^4$C mol^{-1}). The corresponding heat of reaction for an isothermal charging is, from (8·23),

$$\Delta H = zF'\left(\mathscr{E} - T\frac{\mathrm{d}\mathscr{E}}{\mathrm{d}T}\right). \quad (8.24)$$

The importance of this equation is that it provides a method of measuring the heat of reaction of a chemical reaction without resort to calorimetry. One needs only to measure the e.m.f. of a cell and its temperature coefficient. Results for several cells are given in Table 8.2. A positive value for ΔH indicates an endothermic reaction, a negative value an exothermic reaction.

The essential difference between the sort of reversible cell we have discussed and the fuel cell, in which there is at present so much interest, is that the latter is based on a continuous process. The initial chemicals are fed continuously to the cell where they react and the products discharged continuously from it. As in the simple cell, a concentration gradient of the ions involved in the reaction has to be set up between the electrodes. This is usually done by using porous electrodes and feeding the fuels in through them to an intervening electrolyte which is commonly a molten salt. In such a cell more potent chemical reactions can often be used. A cell in which graphite is oxidized to CO_2, for instance, yields a heat of reaction of 390 kJ mol^{-1} (cf. Table 8.2). In principle, a reversible fuel cell converts the whole of the free energy of the chemical reaction into electric power. This is its great advantage over a process in which heat is extracted from a chemical reaction (as in a conventional power station) and then converted into electrical power via a heat engine with its inevitable loss of efficiency.[10]

Table 8.2. Some Reversible Cells*

Reaction	T/K	z	$\mathscr{E}$/V	$\dfrac{d\mathscr{E}}{dT}$/mV K^{-1}	ΔH/kJ mol^{-1}
$Zn + CuSO_4 = Cu + ZnSO_4$	273	2	1·0934	−0·4533	−234·5
$Zn + 2AgCl = 2Ag + ZnCl_2$	273	2	1·0171	−0·2103	−207·1
$Cd + 2AgCl = 2Ag + CdCl_2$	298	2	0·6753	−0·65	−167·6
$Pb + 2AgI = 2Ag + PbI_2$	298	2	0·2135	−0·173	−51·1
$Ag + \frac{1}{2}Hg_2Cl_2 = Hg + AgCl$	298	1	0·0455	+0·338	+5·4
$Pb + Hg_2Cl_2 = 2Hg + PbCl_2$	298	2	0·5356	+0·145	−95·8
$Pb + 2AgCl = 2Ag + PbCl_2$	298	2	0·4900	−0·186	−105·0

* From M. W. Zemansky, *Heat and Thermodynamics*, p. 294, McGraw-Hill, 1957.

8.6. Surface Tension

As our first example of a system with more than two independent variables we shall consider the case of a film in which volume effects are not negligible. The first law becomes

$$dU = T\,dS - p\,dV + \gamma\,dA \tag{8.25}$$

with proper variables S, V, and A. In terms of T, p, and A as independent variables this becomes:

$$\begin{aligned} dU = &\left[T\left(\frac{\partial S}{\partial T}\right)_{p,A} - p\left(\frac{\partial V}{\partial T}\right)_{p,A}\right] dT \\ &+ \left[T\left(\frac{\partial S}{\partial p}\right)_{A,T} - p\left(\frac{\partial V}{\partial p}\right)_{A,T}\right] dp \\ &+ \left[\gamma + T\left(\frac{\partial S}{\partial A}\right)_{T,p} - p\left(\frac{\partial V}{\partial A}\right)_{T,p}\right] dA. \end{aligned} \tag{8.26}$$

[10] For further details see J. Kaye and J. A. Welsh, *Direct Conversion of Heat into Electricity*, Wiley, 1960.

The two $\mathrm{d}T$ terms are easily interpretable. The first is simply the heat capacity at constant p and A. The second is related to the expansion coefficient at constant p and A and represents mechanical work done by the hydrostatic pressure during thermal expansion. Similarly, the second $\mathrm{d}p$ term is related to the compressibility at constant T and A and corresponds to mechanical work done by the hydrostatic pressure due to change of volume brought about by change of pressure. The interpretation of the remaining thermodynamic coefficients is less obvious, and it is helpful to transform them by means of Maxwell relations. These are most conveniently derived by constructing the differential of the appropriate potential by applying a Legendre transformation to $\mathrm{d}U$, and then proceeding by one of the methods used in section 7.3 to derive the fundamental Maxwell relations. We note that all the coefficients in (8.26) have T, p, and A as independent variables so that the potential we require, G', is defined by

$$\begin{aligned} \mathrm{d}G' &= \mathrm{d}U - \mathrm{d}(TS) + \mathrm{d}(pV) \\ &= -S\,\mathrm{d}T + V\,\mathrm{d}p + \gamma\,\mathrm{d}A. \end{aligned} \tag{8.27}$$

Differentiating twice with respect to two variables in both orders and equating, or simply applying the condition that $\mathrm{d}G'$ is exact, we obtain immediately three Maxwell relations:

$$\left.\begin{aligned} \left(\frac{\partial S}{\partial p}\right)_{A,T} &= -\left(\frac{\partial V}{\partial T}\right)_{p,A} \\ \left(\frac{\partial S}{\partial A}\right)_{T,p} &= -\left(\frac{\partial \gamma}{\partial T}\right)_{p,A} \\ \left(\frac{\partial V}{\partial A}\right)_{T,p} &= \left(\frac{\partial \gamma}{\partial p}\right)_{A,T}. \end{aligned}\right\} \tag{8.28}$$

With these we may interpret the remaining differential coefficients in (8.26). The first differential coefficient in the $\mathrm{d}p$ term becomes simply $(\partial V/\partial T)_{p,\mathrm{A}}$ which is related to the cubic expansivity; and the middle term in the $\mathrm{d}A$ group becomes the temperature coefficient of the surface tension. Then, substituting with the isobaric, constant area, heat capacity:

$$T\left(\frac{\partial S}{\partial T}\right)_{p,A} = C_{p,A}^{(T)},$$

the isobaric, constant area, cubic expansivity:

$$\frac{1}{V}\left(\frac{\partial V}{\partial T}\right)_{p,A} = \beta_{p,A},$$

the isothermal, constant area, compressibility:

$$-\frac{1}{V}\left(\frac{\partial V}{\partial p}\right)_{A,T}=\kappa_{A,T},$$

(8.26) becomes:

$$\begin{aligned}\mathrm{d}U=&(C_{p,A}^{(T)}-pV\beta_{p,A})\,\mathrm{d}T\\&+(pV\kappa_{A,T}-TV\beta_{p,A})\,\mathrm{d}p\\&+\left[\gamma-T\left(\frac{\partial\gamma}{\partial T}\right)_{p,A}-p\left(\frac{\partial V}{\partial A}\right)_{T,p}\right]\mathrm{d}A.\end{aligned}\tag{8.29}$$

The significance of the last term, $p(\partial V/\partial A)_{T,p}\,\mathrm{d}A$, is of considerable interest. At first sight one would not expect the volume of liquid to depend on the surface area of the film. The coefficient $(\partial V/\partial A)_{T,p}$ can only be different from zero if the average volume occupied by the molecules near the surface of the film is different from that occupied in the interior. If $(\partial V/\partial A)_{T,p}<0$, then the density of molecules must be greater nearer the surface for as the film is expanded the proportion of matter close to the surface increases and the average density increases. Increasing density near the surface is known as *positive surface adsorption*. In this case, it follows from the Maxwell relation that the surface tension decreases with increasing pressure. The kind of surface adsorption which occurs in a given liquid or solution depends on the nature of the surface forces.

For films at constant temperature and pressure, and neglecting the last term which is usually small, (8.29) becomes

$$\mathrm{d}U_{T,p}=\left[\gamma-T\left(\frac{\partial\gamma}{\partial T}\right)_{p,A}\right]\mathrm{d}A.\tag{8.30}$$

Then the internal energy of the film per unit area is

$$u_{T,p}=\gamma-T\left(\frac{\partial\gamma}{\partial T}\right)_{p,A}.\tag{8.31}$$

u_T is called the *surface energy* of the film. Normally, γ depends very little on p and not at all on A so that (8.31) may be rewritten

▶ $$u_T=\gamma-T\frac{\mathrm{d}\gamma}{\mathrm{d}T}.\tag{8.32}$$

The second term in (8.31) or (8.32) corresponds to exchange of heat with the surroundings during the isothermal change of area. γ

normally decreases with temperature (vanishing at the critical temperature) so that on stretching, heat is absorbed and the surface energy of a film is always greater than the mechanical work put in to create it. It should be noted that (8.32) is an analogue of the Gibbs–Helmholtz equation (see section 7.1) with γ as the Helmholtz free energy per unit area.

8.7. Piezoelectricity

We take one further example with three degrees of freedom: the elastic behaviour of a material subjected to an electric field.

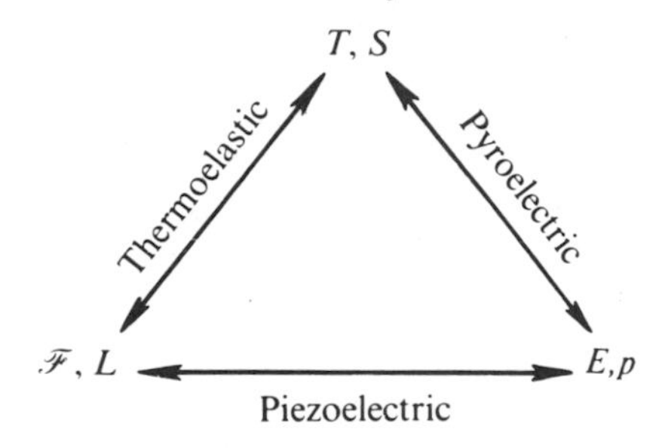

Fig. 8.4. The relationship between the thermoelastic, pyroelectric, and piezoelectric effects.

When an elastic substance is stretched isothermally heat is exchanged with the surroundings and its entropy changes. If the process is performed isentropically, its temperature changes. Such effects may be called *thermoelastic*. Similarly, changing the polarization of a dielectric causes changes in T or S. Such effects may be called *pyroelectric*. If a system in an electric field suffers changes in polarization when the tension is changed, or conversely changes in tension when the polarization is changed it is said to show *piezoelectric* effects. The relationship between the various parameters is shown in Fig. 8.4.

The coordinates for an isotropic material subjected to tension and electric field are T, S; $\mathcal{F}$, L; E, p. The first law becomes

$$\mathrm{d}U = T\,\mathrm{d}S + \mathcal{F}\,\mathrm{d}L + E\,\mathrm{d}p. \tag{8.33}$$

Since we are normally interested in changes at constant temperature, constant field, or constant tension, it is convenient to construct the potential function with these three quantities as its proper variables, $G(T, E, \mathcal{F})$, which is sometimes called the piezoelectric Gibbs function. We have to change the proper variables in each of the terms of U. Following the Legendre method,

$$\begin{aligned} dG &= dU - d(TS + \mathscr{F}L + Ep) \\ &= -S\,dT - L\,d\mathscr{F} - p\,dE. \end{aligned} \tag{8.34}$$

Applying the condition that this is an exact differential we obtain three Maxwell equations:

$$\left.\begin{aligned} &\text{thermoelastic:} && \left(\frac{\partial S}{\partial \mathscr{F}}\right)_{E,T} = \left(\frac{\partial L}{\partial T}\right)_{\mathscr{F},E} \\ &\text{pyroelectric:} && \left(\frac{\partial S}{\partial E}\right)_{T,\mathscr{F}} = \left(\frac{\partial p}{\partial T}\right)_{\mathscr{F},E} \\ &\text{piezoelectric:} && \left(\frac{\partial L}{\partial E}\right)_{T,\mathscr{F}} = \left(\frac{\partial p}{\partial \mathscr{F}}\right)_{T,E}. \end{aligned}\right\} \tag{8.35}$$

The first two coefficients in (8.35) are simply related to the heat absorbed in isothermal changes of field or tension, and are similar to the expressions which would be obtained by analysing the simple two-parameter systems where the electrical or tension effects are negligible. They may be simplified:

$$C^{(\mathscr{F})}_{E,T} = T\left(\frac{\partial S}{\partial \mathscr{F}}\right)_{E,T} = T\left(\frac{\partial L}{\partial T}\right)_{\mathscr{F},E} = LT\alpha_{\mathscr{F},E}$$

where $\alpha_{\mathscr{F},E}$ is the linear expansivity at constant tension and constant field, and

$$C^{(E)}_{T,\mathscr{F}} = T\left(\frac{\partial S}{\partial E}\right)_{T,\mathscr{F}} = T\left(\frac{\partial p}{\partial T}\right)_{\mathscr{F},E} = ET\left(\frac{\partial \chi_e}{\partial T}\right)_{\mathscr{F},E}$$

where χ_e is the electrical susceptibility, which is normally independent of field but may depend on temperature. The piezoelectric coefficient in (8.35) is a measure of the extent to which changing the field alters the length of the material. The equation shows that this is related to the sensitivity of the polarizability to the applied force.

Unfortunately, all materials in which the piezoelectric effect is different from zero have a low crystal symmetry, and are highly anisotropic, so that the detailed analysis becomes much more complex than that we have given here.[11] Piezoelectric effects are of great technical importance, however. For example, barium titanate, in which the effect is large, is used in many kinds of electromechanical transducers, and the mechanical vibration of thin quartz plates is used to stabilize the frequency of oscillators.

[11] See W. G. Cady, *Piezoelectricity*, McGraw-Hill, 1946.

8.8. The Magnetocaloric Effect

When the magnetization of a material is changed isothermally, heat is usually exchanged with the surroundings. If the change in magnetization is performed under adiabatic conditions, the temperature changes. Such interdependence of the thermal and magnetic properties is known as the *magnetocaloric effect*. At low temperatures, the effect may become very large and is of considerable importance in providing the basis of the method of obtaining temperatures below 1 K known as *adiabatic demagnetization*. In our analysis, we shall ignore work by hydrostatic pressure which is normally negligible if the magnetic material is a solid. The treatment may easily be modified by the inclusion of the extra variables for a system with three effective degrees of freedom such as a paramagnetic gas.

Neglecting other than magnetic work, and assuming the material to be isotropic, the first law becomes[12]

$$dU = T\,dS + B\,dm.^{13} \tag{8.36}$$

The other potential functions follow from this by Legendre transformation and all the results derived for the simple fluid subjected to hydrostatic pressure may be taken over for the magnetic system by the substitutions $-p \rightarrow B$, $V \rightarrow m$. Choosing B and T as our independent variables the heat *absorbed* in a reversible isothermal change of magnetization is

$$C_T^{(B)} = \frac{đQ_T}{dB} = T\left(\frac{\partial S}{\partial B}\right)_T = T\left(\frac{\partial m}{\partial T}\right)_B \tag{8.37}$$

[12] For some remarks about the somewhat vexed question of magnetic energy see the Appendix.

[13] It is important to remember that here and in the subsequent discussion B is the induction *in the absence of the specimen*. If the material is weakly magnetic ($\chi \ll 1$), internal and applied fields will be approximately equal, but if it is strongly magnetic the internal field may be much smaller than the applied. This difference is treated by introducing the idea of a *demagnetizing factor n*, defined by

$$H_{\text{int}} = H_{\text{ext}} - nM$$

in which n ranges from 0 to 1 for different shaped bodies. The apparent susceptibility becomes

$$\chi_{\text{app}} = \chi/(1 + n\chi).$$

See B. I. Bleaney and B. Bleaney, *Electricity and Magnetism*, p. 141, Oxford, 1965.

where the last step involves the use of a Maxwell relation. In a reversible adiabatic change of magnetization the change in temperature is

$$\left(\frac{\partial T}{\partial B}\right)_S = -\left(\frac{\partial T}{\partial S}\right)_B \left(\frac{\partial S}{\partial B}\right)_T = -\frac{T}{C_B}\left(\frac{\partial m}{\partial T}\right)_B \tag{8.38}$$

where C_B is the heat capacity at constant induction. If the susceptibility, $\chi = m/VH_{\text{int}}$, is not too large, we may neglect demagnetizing effects[13] and put $H_{\text{int}} \simeq B/\mu_0$, where B is the applied induction in the absence of the material. Then, in terms of χ, equations (8.37) and (8.38) become

$$T\left(\frac{\partial S}{\partial B}\right)_T = \frac{TVB}{\mu_0}\left(\frac{\partial \chi}{\partial T}\right)_B \tag{8.39}$$

and

▶ $$\left(\frac{\partial T}{\partial B}\right)_S = -\frac{TVB}{\mu_0 C_B}\left(\frac{\partial \chi}{\partial T}\right)_B. \tag{8.40}$$

It should be noted that both (8.39) and (8.40) contain the temperature derivative of the susceptibility, so that materials in which the susceptibility does not vary with temperature show no magnetothermal effects. Such is the case with simple diamagnetism where the magnetic response results from the perturbation by the applied field of the electronic eigenstates in the atoms. These are essentially unaffected by temperature. Paramagnetism on the other hand always results from the presence of microscopic magnetic dipoles in the material which may be aligned by the application of a field. Thermal motions tend to disalign the dipoles so that the extent of the alignment decreases with increasing temperature. Thus $(\partial \chi_m/\partial T)_B$ is always negative and, according to (8.39), heat is evolved in an isothermal magnetization. This is in agreement with what we should expect from the connection between entropy and order, for the magnetization of a paramagnet increases the magnetic ordering and thus decreases the magnetic contribution to the entropy. In an isothermal change, heat will therefore be evolved. The dependence of the susceptibility of a paramagnet on temperature increases rapidly as the temperature is reduced towards the point where spontaneous magnetic ordering sets in. It is therefore in this region that strong magnetothermal effects are to be expected. However, they are only important at low temperatures where specific heats are generally small and relatively large changes of temperature may be produced. Until the development of the dilution refrigerator,[14] the magnetocaloric effect provided the only useful method of obtaining temperatures below 0·3 K. In this context it has been and still is of great importance as an experimental tool.

8.8.1. Cooling by Adiabatic Demagnetization

Figure 8.5 shows the variation of entropy with temperature and magnetic field for a typical paramagnetic salt. In zero field the fall in entropy at the *Curie temperature* T_c corresponds to the onset of spontaneous ordering. At higher temperatures the entropy may always be reduced by applying a magnetic field and so increasing the magnetic order. The process of cooling the salt is illustrated in the figure. The salt is first magnetized by applying a field of induction B_1 at an initial temperature T_1 which is usually obtained by evaporating liquid ^{4}He or liquid ^{3}He under reduced pressure.[15] The heat evolved during magnetization is conducted away to the helium bath, and the entropy falls, the salt going from state a to state b. The salt is then isolated thermally and demagnetized. If the demagnetization is performed sufficiently slowly, the process is reversible, the entropy will remain constant, and the temperature falls. If the field is reduced to zero, the final state of the salt will be

[14] The dilution refrigerator uses the heat of dilution of a solution of ^{3}He in ^{4}He and is capable of achieving temperatures below 20 mK. For a simple description, see D. Fishlock, 'A Superfluid Refrigerator', *New Scientist*, **29**, 755, 1966. For a more detailed discussion, see, H. E. Hall, P. J. Ford, and K. Thompson, *Cryogenics*, **6**, 80, 1966. For lower temperatures adiabatic demagnetization using the magnetocaloric effect is still the only method available.

[15] By evaporating liquid ^{4}He under reduced pressure, a temperature of about 1 K may be obtained fairly easily. With the lighter isotope ^{3}He, the temperature may be reduced to about 0·3 K. There are two reasons for the difference. First, ^{3}He, being lighter than ^{4}He has a greater zero point energy and hence a higher vapour pressure. It is therefore easier to reduce the temperature by pumping on it. Second, below 2·18 K, ^{4}He becomes superfluid. In this state it spreads over all accessible surfaces. In particular, it spreads from the liquid along the container walls and along the pumping tubes to warmer regions where it evaporates, rapidly. This reduces the efficiency of the pumping, and also, some of the vapour recondenses on the liquid, transporting its latent heat in the process. Both these processes oppose the reduction of the temperature. Liquid helium will be discussed in more detail in section 10.9.2.

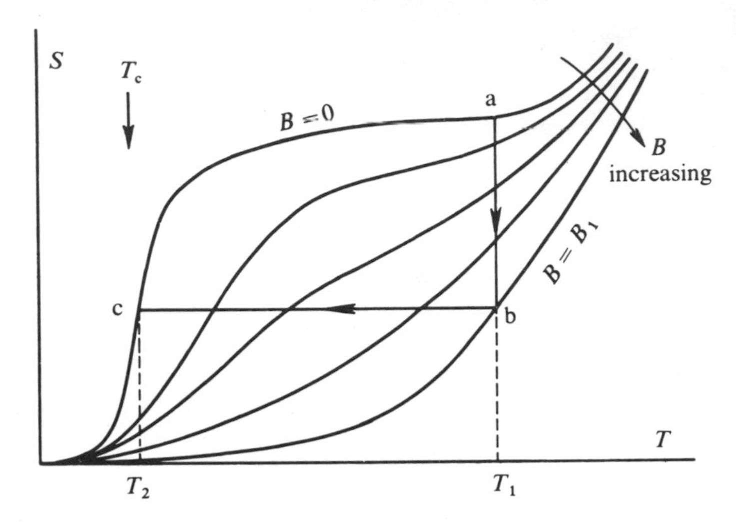

Fig. 8 . 5. The entropy of a paramagnetic salt as a function of temperature and magnetic induction.

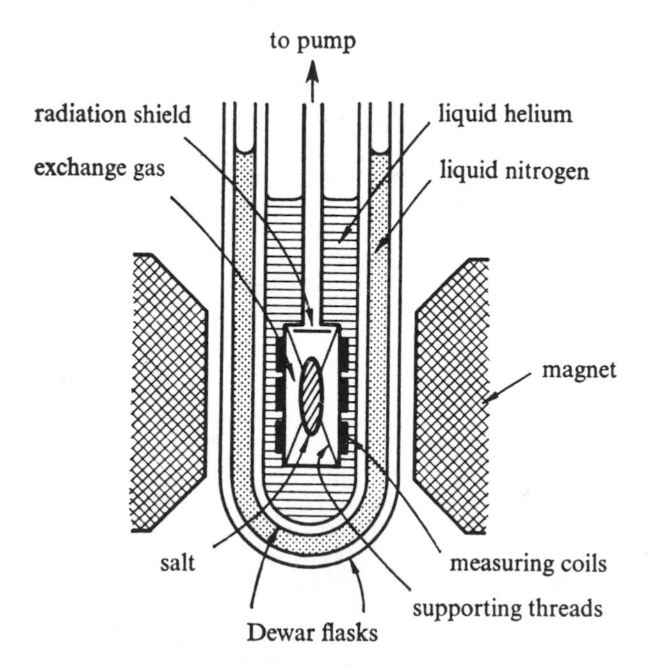

Fig. 8 . 6. A typical arrangement for adiabatic demagnetization experiments.

at c, with temperature T_2. Clearly, the lowest temperature to which the salt can be cooled by demagnetization is effectively the Curie temperature.

The experimental arrangement for adiabatic demagnetization is illustrated in Fig. 8.6. The salt is suspended in a chamber which is immersed in the helium bath which produces the initial cooling. During the isothermal magnetization, thermal contact with the bath is provided by helium 'exchange' gas in the chamber. After magnetization the gas is removed, effectively isolating the sample and the field is then reduced to zero.

The microscopic significance of the adiabatic demagnetization is illustrated in Fig. 8.7. In zero field (a), the energy levels of the microscopic dipoles are close together, their separation being determined by the strength of the interaction between neighbouring dipoles and between the dipoles and the lattice. These different levels correspond to different orientations. If the separation is small in comparison with kT at the initial temperature, the levels will be

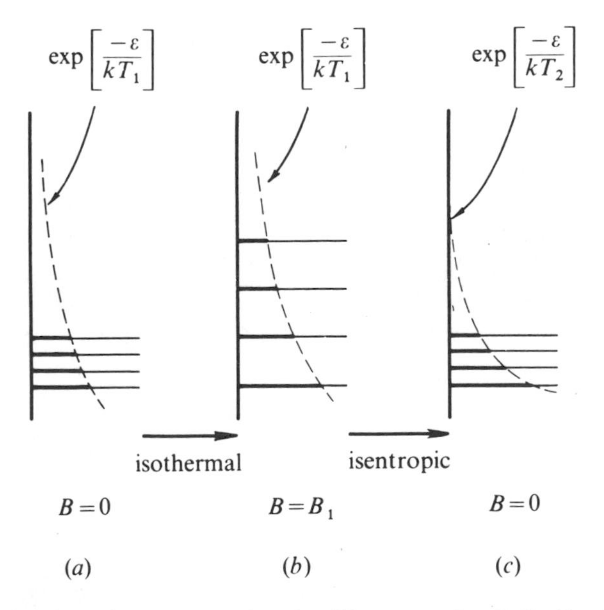

Fig. 8.7. The microscopic significance of adiabatic demagnetization. Energy is plotted vertically and occupation of the levels indicated by the length of the heavy horizontal lines.

nearly equally populated.[16] Application of a field (b) causes the levels to separate. Transitions then take place between the levels in which the magnetic subsystem loses energy to the surroundings (to the helium bath via the crystal lattice and the exchange gas) and a new distribution among the levels is established, characteristic of the same temperature, but with the levels differently populated because of their changed energies. Under the adiabatic change, transitions do not occur (this is the microscopic significance of 'adiabatic') so that when the levels return to their original separations on removal of the field (c), the populations are characteristic of a much lower temperature.

Spontaneous magnetic ordering occurs when the thermal energies become smaller than the energy differences between the various possible orientations of the dipoles in the absence of an applied field. These energy differences result from the interactions of the dipoles with one another and with the lattice. The stronger they are, the higher the Curie temperature will be and the higher the temperature reached by demagnetization. This is one of the reasons why the paramagnetic salts used for adiabatic demagnetization are usually chemically complex. By 'diluting' the active magnetic ions so that their mean separation is greater, their interaction energy is reduced. This helps to lower the Curie temperature; but it also means that the magnetic entropy of the salt is smaller so that if heat has to be absorbed from elsewhere it becomes a less powerful cooling agent. This is of importance in experiments where the salt is used as a means of cooling some other experimental system. In such a case, the entropy of the system must be added to that of the salt in calculating the effect of the isentropic process.

When the salt is used as a cooling agent, the condition of reversibility in the demagnetization process becomes much more stringent. Reversibility requires thermal equilibrium at all times throughout the salt and all that is in thermal contact with it. When the salt alone is being cooled, the only relaxation time involved is that for the spin system (on which the magnetic field acts directly) to reach equilibrium with the crystal lattice in which it is situated. This time varies rapidly with temperature, but for magnetically dilute salts is of the order of 1 s at 1·5 K. When another system has to

[16] The relative populations of the levels will vary as exp $(-\varepsilon/kT)$ where ε is the energy of the level and k is the Boltzmann constant. See F. Mandl, *Statistical Physics*, pp. 53–7, Wiley, 1971; or C. Kittel, *Thermal Physics*, chapter 6, Wiley, 1969.

be cooled, however, the limiting factor is usually the thermal contact between the salt and the rest of the system. At very low temperatures, the boundaries between materials offer a high thermal resistance[17] and it may take many hours to achieve thermal equilibrium.

Thermodynamics cannot, of course, predict the behaviour of a particular substance unless sufficient information is given about its properties.[18] To use equation (8.38) to calculate magnetic cooling it is enough to know its equation of state, $M=M(B,T)$, and its specific heat in zero field over the relevant temperature range. We may illustrate the thermodynamics by taking a highly simplified model.

Sufficiently far above its Curie point, the susceptibility of a paramagnetic material is essentially independent of magnetic field and obeys Curie's law, $\chi=a/T$. This takes no account of the interactions which bring about spontaneous ordering at the Curie temperature close to which the susceptibility will rise sharply. A better approximation is the Curie–Weiss law[19]

$$\chi=\frac{a}{T-T_c} \tag{8.41}$$

where T_c is the Curie temperature. This is, in reality, still a poor approximation close to the Curie point where χ becomes large but not infinite and is no longer independent of B. However, for our model, we will take the susceptibility to obey the Curie–Weiss law.

[17] Heat is carried by phonons and, in conductors, by the electrons. In pure materials at very low temperatures, the thermal resistance is dominated by scattering at the surface of the sample and at crystallite boundaries. Although the total conductivity may be low, the mean free paths of the phonons and electrons contributing to it may be very large. (In high quality crystalline alumina below 1 K the mean free path of the phonons may be many centimetres.) The extra scattering introduced at crystallite boundaries or contacts between materials may therefore be *relatively* enormous. For a discussion of the mechanisms contributing to thermal resistance see H. M Rosenberg, *Low Temperature Solid State Physics*, chapter 3, Oxford University Press, 1963.

[18] The information may be given in many forms. One could, for example, start with a knowledge of the microscopic structure and use the techniques of statistical mechanics to derive the thermodynamic parameters. For a simple illustration see F. Mandl, *Statistical Physics*, section 5.6, Wiley, 1971.

[19] For the derivation of these formulae see C. Kittel, *Introduction to Solid State Physics*, 502–3, and 529–30, Wiley, 1971.

We first calculate $C_B(B, T)$.

$$\left(\frac{\partial C_B}{\partial B}\right)_T = \left.\frac{\partial}{\partial B}\right)_T \left[T\left(\frac{\partial S}{\partial T}\right)_B\right] = T\,\frac{\partial^2 S}{\partial B\,\partial T}$$

$$= T\,\frac{\partial^2 S}{\partial T\,\partial B} = T\left.\frac{\partial}{\partial T}\right)_B \left(\frac{\partial S}{\partial B}\right)_T$$

$$= T\left.\frac{\partial}{\partial T}\right)_B \left(\frac{\partial m}{\partial T}\right)_B \qquad \text{(by a Maxwell)}$$

$$= T\left(\frac{\partial^2 m}{\partial T^2}\right)_B$$

$$= \frac{TVB}{\mu_0}\left(\frac{\partial^2 \chi}{\partial T^2}\right)_B \tag{8.42}$$

where the final substitution again assumes demagnetizing effects to be negligible. But, from (8.41)

$$\left(\frac{\partial^2 \chi}{\partial T^2}\right)_B = \frac{2a}{(T-T_c)^3}. \tag{8.43}$$

Substituting (8.43) in (8.42) we obtain

$$\left(\frac{\partial C_B}{\partial B}\right)_B = \frac{2aTVB}{\mu_0\,(T-T_c)^3},$$

and integrating

$$C_B(B, T) = C_B(0, T) = \int_0^B \frac{2aTVB}{\mu_0(T-T_c)^3}\,\mathrm{d}B$$

$$= C_B(0, T) + \frac{aTVB^2}{\mu_0(T-T_c)^3}. \tag{8.44}$$

The second term is associated with the change of magnetic order brought about by the applied field. The first term contains all other contributions to the heat capacity. These are

(*a*) the contributions of the lattice containing the magnetic ions and of any material cooled by the salt. These are often very small at low temperatures, and we shall neglect them.

(*b*) the contribution deriving from the spontaneous change in magnetic order which takes place in zero field near the Curie point.[20] This is not small but the contribution is peaked close

[20] See the discussion of entropy and order in section 5.6.2.

to the Curie temperature and at higher temperatures is relatively small.

Provided then that we do not come too close to the Curie point we may neglect $C_B(0, T)$ and take C_B as:

$$C_B(B, T) = \frac{aTVB^2}{(T-T_c)^3}. \tag{8.45}$$

Substituting (8.42) and (8.45) in the expression for the magnetic cooling, equation (8.40),

$$\left(\frac{\partial T}{\partial B}\right)_S = \frac{T-T_c}{B},$$

and integrating

$$\frac{T_1-T_c}{T_2-T_c} = \frac{B_1}{B_2}. \tag{8.46}$$

In this approximation, demagnetizing to zero field cools the salt to the Curie temperature. In practice, the temperature does not drop as much as this because the approximations we have made cease to be good near the Curie point.

From the foregoing discussion it is clear that the temperature reached in adiabatic demagnetization is the result of a compromise. The salt must be sufficiently dilute magnetically for the interactions between the magnetic atoms to be small and the Curie temperature low. On the other hand, if the salt is too dilute, the entropy change associated with magnetization becomes small in comparison with the entropy of the rest of the system, and this again restricts the cooling which can be obtained.

To achieve the lowest temperatures, extremely sophisticated experimental techniques have to be used.[21] With cerium magnesium nitrate, it is possible to reach a few millikelvins. For lower temperatures, weaker magnetic systems have to be used. Adiabatic demagnetization of *nuclear* magnetic moments, previously cooled by an ordinary paramagnetic salt, have produced nuclear spin temperatures of about 1 μK, but, of course, the lattice was not cooled to this temperature.[22]

8.8.2. The Measurement of Temperatures below 1°K

Having achieved the very low temperatures possible by adiabatic

[21] O. E. Vilches and J. C. Wheatley, *Rev. Sci. Instrum.*, **37**, 819, 1966.

[22] N. Kurti, *Physics Today*, **13**, No. 10, p. 26, 1960.

demagnetization there remains the problem of measuring them. Down to 1 K, direct gas thermometry is possible so that the problem there is simply one of thermometer calibration. Below 1 K some different method has to be used to determine the absolute temperature.

The simplest method is to use the paramagnetic salt itself as a thermometer, taking its susceptibility as the thermometric property. From this one first defines an empirical temperature scale. It is most convenient to do this in such a way that at high temperatures, above 1 K, the empirical and thermodynamic temperatures coincide. If the salt is used to cool to well below 1 K, then, in the helium range, it will obey Curie's law accurately. Thus, we ensure that the scales coincide above 1 K if we define the empirical temperature via Curie's law:

$$\chi_m = a/T^*. \tag{8.47}$$

T^* is called the *magnetic temperature,* and the constant a is determined by measurements in the helium range. T and T^* will then diverge at lower temperatures and their relationship may be established by the following procedure, which is illustrated in Fig. 8.8. It uses the fact that reversible adiabatic changes are isentropic.

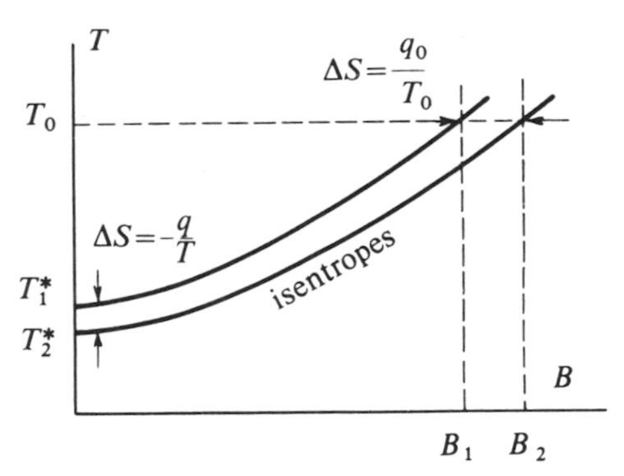

Fig. 8.8. The calibration of a susceptibility thermometer.

1. Magnetize the salt isothermally in a field of induction B_1 at T_0 in the helium range.
2. Demagnetize adiabatically to zero field and observe the magnetic temperature $T^* = a/\chi_1$.
3. Magnetize the salt isothermally in a field of induction B_2, slightly greater than B_1, at T_0.
4. Demagnetize adiabatically to zero field and observe the new magnetic temperature T_2^*.

5. Heat the salt (with $B=0$) from T_2^* to T_1^* by adding a known amount of heat, q.
6. Measure the difference in entropy between the two magnetized states, $S(B_1, T_0)-S(B_2, T_0)$.

Since the demagnetizations are isentropic, the entropy change in 5 must be equal to the entropy difference measured in 6. By making B_1 and B_2 close together, T_1^* and T_2^* will be close and we may find their mean *thermodynamic* temperature T. Then the entropy condition gives

$$q=T\{S(B_1, T_0)-S(B_2, T_0)\} \tag{8.48}$$

from which T may be found.

The susceptibility is usually obtained by measuring the inductance of a coil surrounding the salt. The heat required to raise the temperature from T_2^* to T_1^* may be measured either by heating the salt with an electrical heater, or by using its natural hysteresis to cause power absorption when an alternating field is applied. (The imaginary part of the susceptibility, corresponds to the losses in the salt, and may be measured at the same time as the real part.) The latter

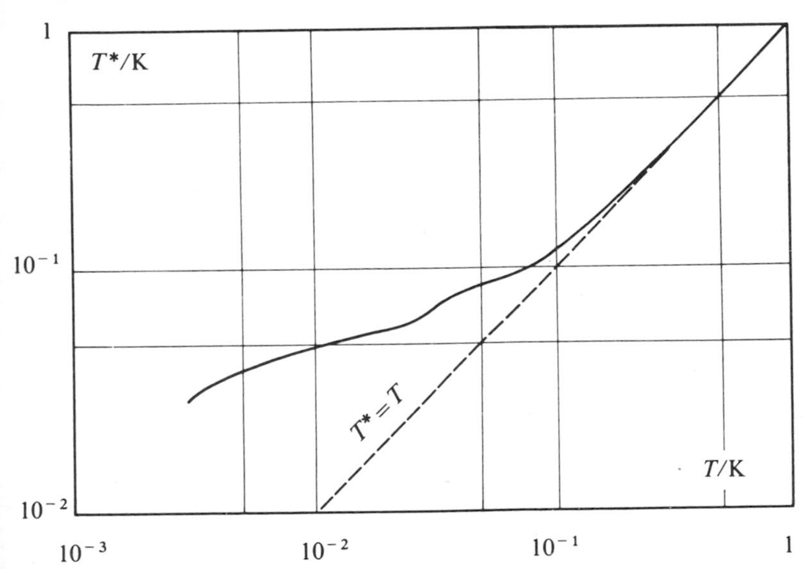

Fig. 8.9. The relationship between the magnetic temperature and the Kelvin temperature for chromium potassium alum.[23]

[23] B. Bleaney, *Proc. Roy. Soc.*, **A204**, 216, 1950; and D. de Klerk, M. J. Steenland and C. J. Gorter, *Physica*, **15**, 649, 1949.

method has the advantage that the heating is applied uniformly through the salt so that temperature differences are less likely to occur. The entropy difference, $S(B_1, T_0) - S(B_2, T_0)$, may be measured directly by determining the heat of magnetization, or calculated from (8.47) using the relation

$$\left(\frac{\partial S}{\partial B}\right)_T = VB\left(\frac{\partial \chi}{\partial T}\right)_B.$$

A typical relationship between T and T^* is shown in Fig. 8.9.

8.9. Thermal Radiation

8.9.1. The Basic Properties of Thermal Radiation

All bodies emit electromagnetic radiation by virtue of their temperature. The character of this radiation depends on how hot the body is and on the nature of its surface. At room temperature, most of the radiated energy is in the far infra-red, but at higher temperatures the region of strongest emission shifts to shorter wavelengths. At 6000 K, corresponding to the optical temperature of the sun, it lies in the visible part of the spectrum.

Thermal radiation has the usual properties of electromagnetic waves. It has the velocity of light and can be reflected, refracted and suffers diffraction appropriate to its wavelength. It carries energy, and, when absorbed or reflected, exerts a pressure. These properties are a straightforward consequence of electromagnetic theory;[24] but we shall avoid using electromagnetic theory here. Instead, we shall think of the radiation as a gas of photons and draw on some of the elementary results of the kinetic theory of gases.[25]

If *isotropic* radiation is trapped in a vessel with perfectly reflecting walls, then the photons of which the radiation consists move in random directions within the container and rebound elastically from its walls. The situation is analogous to that of an ordinary gas contained in a similar vessel. For this, the kinetic theory gives the number of molecules striking unit area of the wall per second,

$$\mathcal{N} = \tfrac{1}{4} n\bar{c}$$

where n is the number of molecules per unit volume and $\bar{c}$ their

[24] See, for example, B. I. and B. Bleaney, *Electricity and Magnetism*, Oxford University Press, 1965.

[25] The results we shall use are derived in B. H. Flowers and E. Mendoza, *Properties of Matter*, Wiley, 1970.

mean velocity. For our photon gas, all the photons move with the same velocity, namely the velocity of light, so that the number of photons hitting unit area of the wall per second is simply

$$\mathcal{N} = \tfrac{1}{4}nc.$$

If the average photon energy is η, then the energy incident per second per unit area of the container wall is

$$\mathcal{P} = \tfrac{1}{4}n\eta c = \tfrac{1}{4}uc \qquad \textit{(isotropic radiation)} \quad (8.49)$$

where u is the *energy density* of the radiation. If a small hole of area δA is cut in the wall of the container, the energy escaping per second is[26]

$$\delta\mathcal{P} = \tfrac{1}{4}uc\,\delta A \qquad \textit{(isotropic radiation)} \quad (8.50)$$

The pressure due to isotropic radiation may be derived in a similar way. The pressure exerted by a gas is

$$p = \tfrac{1}{3}\rho\overline{c^2} \rightarrow \tfrac{1}{3}\rho c^2$$

where ρ is the mass density and $\overline{c^2}$ becomes c^2 for our photon gas. Using the Einstein mass-energy relation,[27] $u = \rho c^2$, giving

$$p = \tfrac{1}{3}u \qquad \textit{(isotropic radiation)} \quad (8.51)$$

The macroscopic mechanism by which this force is communicated to the surface is shown in Fig. 8.10.

It should be noted that in both these results u is the *total* energy density and includes radiation both approaching and receding from the surface. If the radiation is absorbed or partially absorbed, (8.51) still holds *provided that any reflected radiation is also isotropic.* This follows immediately from momentum considerations, for any photon which is absorbed exchanges half as much momentum normal to the surface as it would if reflected; but then it does not contribute to u after impact. u is thus reduced in proportion to the

[26] For this, or the equivalent kinetic result, to hold, the efflux of radiation or of molecules has to be sufficiently restricted so that the conditions within the container are always quasistatic. This will ensure that the fluxes remain isotropic. In the case of the gas, it is sufficient if the linear dimension of the hole is much smaller than a mean free path of the molecules. For radiation, the requirement is equivalent to making the time required for the greater part of the radiation to escape very much longer than the time taken for the radiation to redistribute itself throughout the container. This amounts to satisfying the inequality: (area of hole)$^3 \ll$ (volume of container)^2.

[27] See, A. P. French, *Special Relativity*, Nelson, 1968.

momentum exchange. In general, cases where the radiation is not isotropic have to be considered individually. The only important case is that of radiation incident normally. For this, the corresponding results are

$$\mathscr{P} = u'c \qquad (\textit{normal incidence}) \quad (8.52)$$

where u' *only includes radiation approaching the surface* and

$$p = u \qquad (\textit{normal incidence}) \quad (8.53)$$

where u *is the total energy density provided that any reflection is specular.*

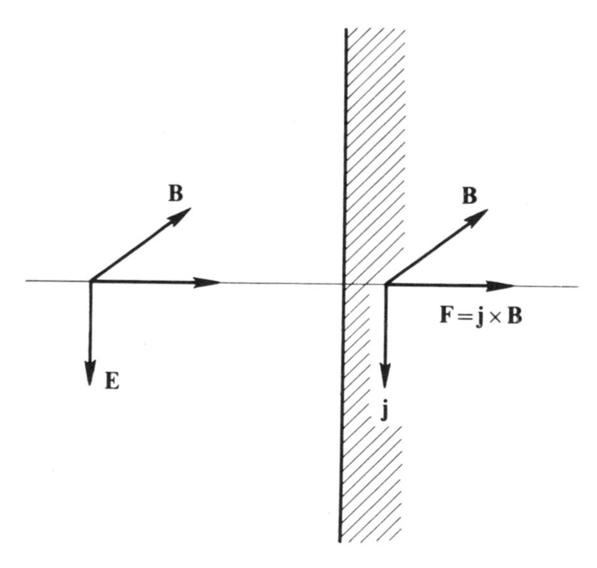

Fig. 8 . 10. The macroscopic mechanism by which radiation exerts a pressure.

The electric and magnetic components of the electromagnetic field are perpendicular to each other and to the direction of propagation. The electric fields induce currents **j** in the material which interact with the magnetic field to give a force in the direction of propagation.

Because thermal radiation transports energy, all bodies continually lose energy to and receive energy from their surroundings. If a body emits more radiation than it absorbs then it suffers a net loss of energy,[28] and if it reaches equilibrium, then the equilibrium is

[28] We are assuming that radiation is the only means by which energy is exchanged.

dynamic. This is the *theory of exchanges* first put forward by Prévost after noting that objects initially in thermal equilibrium with their surroundings became cooler if a cold body was placed nearby.

Before proceeding to the detailed discussion of the properties of radiation it is convenient to define three quantities which we shall need to use. We have already defined u as the total energy density due to radiation. It will be necessary to discuss how this energy is distributed with wavelength. We therefore define the

spectral energy density u_λ, such that $u_\lambda \, d\lambda$ is the energy density contained in radiation in the wavelength range between λ and $\lambda + d\lambda$. We also define the

spectral absorptivity of a surface, α_λ, which is the fraction of the incident radiation at wavelength λ which is absorbed; and the

spectral emissive power of a surface, e_λ, such that $e_\lambda \, d\lambda$ is the energy emitted per unit area per second by the surface in radiation in the wavelength range from λ to $\lambda + d\lambda$.[29]

8.9.2. Equilibrium Radiation

An *equal temperature enclosure* is one in which all parts of the walls are at the same temperature. Some short time after we have set up such an enclosure we expect that the radiation it contains will have reached a state of equilibrium with the walls surrounding it. That is, if we measure its spectral energy density at subsequent times we will always obtain the same result.

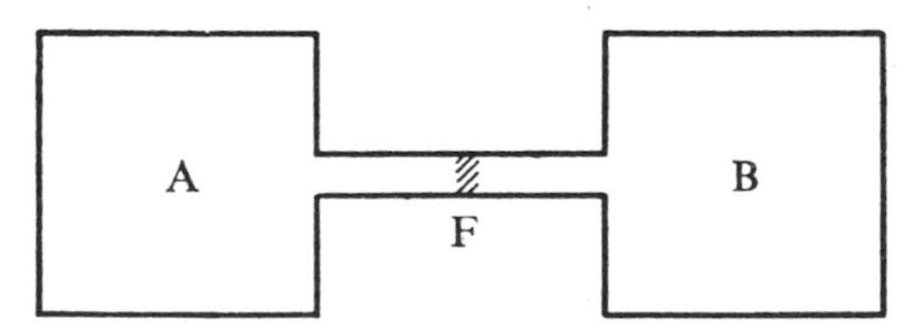

Fig. 8.11. The proof that the spectral energy density of equilibrium radiation is a function of temperature and wavelength only.

We now consider two equal temperature enclosures A and B which are initially at the same temperature. We then join them by a

[29] It is helpful to note the convention whereby a Greek letter is used for a dimensionless ratio, whereas a Roman letter is used for a quantity which has dimensions. By analogy with absorptivity, we shall later define emissivity ε_λ as the ratio of the emissive power of a surface to that of a perfect emitter.

narrow tube in which we insert a filter F which passes a narrow band of wavelengths centred at λ[30] (Fig. 8.11). Now suppose that the spectral energy density at λ in A is greater than that in B:

$$u_\lambda^A > u_\lambda^B.$$

Then there will be a net flow of energy from A to B. This will cause the temperature of B to rise and that of A to fall. But the divergence of the temperatures of two bodies placed in thermal contact (by whatever means) is forbidden by the second law. It involves a universal decrease in entropy.[31] Thus, we must have

$$u_\lambda^A = u_\lambda^B. \tag{8.54}$$

This means that the nature of the radiation is independent of the nature of the walls of the equal temperature enclosures. It must, by the same token be isotropic. It can only depend on temperature.

We have thus shown that for radiation *in equilibrium* with the walls of an equal temperature enclosure

▶ $$u_\lambda = f(\lambda, T). \tag{8.55}$$

By its universal nature, this radiation proves to be very important and is variously known as *equilibrium radiation*, or *full radiation*, or for a reason which will later become apparent, *black body radiation*. Its importance lies in the fact that we are able to relate this radiation, which occurs in an equilibrium situation, to the radiation emitted by bodies which are *not* in equilibrium with their surroundings.

It is worth pointing out that if an equal temperature enclosure contains bodies which are allowed to reach thermal equilibrium, then the same arguments may again be applied to show that the radiation everywhere is still isotropic and of a nature depending on temperature only. An obvious consequence of this is that in such a situation it is impossible to detect the presence of any objects by measurements of the radiation. In all directions one would observe a uniform brightness. Conversely, the print on this page is only visible because the radiation reaching it is very far from being in

[30] We must assume that the tube is narrow enough to restrict the energy flux from one cavity to the other sufficiently for the changes occurring in each cavity to be quasistatic. In each cavity, the radiation then remains in equilibrium with the cavity at all times.

[31] Alternatively, we could warm B slightly without reversing the flow of energy thus contriving the situation where energy flows spontaneously from colder to hotter, again violating the second law.

equilibrium with it. There must be a source of high temperature radiation which illuminates it and allows the print to be distinguished by virtue of its different reflecting properties.

8.9.3. Kirchhoff's Law

If equilibrium radiation is independent of the nature of the walls of an equal temperature enclosure, then the walls themselves must have certain properties which preserve the nature of the radiation interacting with them. In particular, if a surface absorbs a certain wavelength strongly it must also emit it strongly. This is expressed more precisely in Kirchhoff's law.

Consider any body inside an equal temperature enclosure and in equilibrium. The radiation is isotropic so that the energy falling on unit area per second due to radiation with wavelengths between λ and $\lambda+\mathrm{d}\lambda$ is

$$\tfrac{1}{4}cu_\lambda\,\mathrm{d}\lambda.$$

Of this, the surface absorbs a fraction α_λ and reflects the rest. But the quality of the radiation is only preserved if the surface also emits as much radiation as it absorbs. Hence we must have

$$e_\lambda\,\mathrm{d}\lambda=\alpha_\lambda\tfrac{1}{4}cu_\lambda\,\mathrm{d}\lambda.$$

But according to (8.55), u_λ is a function of λ and T only. Hence

▶ $$\frac{e_\lambda}{\alpha_\lambda}=\tfrac{1}{4}cu_\lambda=g\,(\lambda, T) \qquad (8.56)$$

where g is a universal function of wavelength and temperature only. This result is *Kirchhoff's Law*: *The ratio of the spectral emissive power to the spectral absorptivity for all bodies is a universal function of wavelength and temperature only.*

It is now convenient to define a *black body* as one which absorbs all the radiation incident on it. Thus, for a black body, $\alpha_\lambda=1$. (If the absorptivity is constant with wavelength but less than unity, the body is *grey* while if α_λ varies with wavelength we say iti *coloured*.) For a black body (8.56) becomes

$$e_\lambda=\tfrac{1}{4}cu_\lambda. \qquad (8.57)$$

Since c is a constant we see that the radiation emitted by a black body has a wavelength and temperature dependence which is identical to that of equilibrium radiation. This is why equilibrium radiation is also called *black body radiation*.

Being a perfect absorber, a black body is, by Kirchhoff's law, also the best possible emitter. This allows us to define the *spectral emissivity* ε_λ of a surface as the ratio of its emissive power to that of a black body. It follows from Kirchhoff's law that $\varepsilon_\lambda = \alpha_\lambda$.

There are many simple illustrations of Kirchhoff's law. We may explain why there is more likely to be a frost on a clear night than on a cloudy one: Interstellar space behaves like a black body with a temperature of a few kelvins. On a clear night, the earth's surface radiates into space but very little energy is incident from space. There is a large net radiative loss and the surface temperature falls (especially if there is no wind to promote warming by the air near the ground). Water, however, absorbs strongly in the infra-red, so that on a cloudy night, not only do the clouds absorb the radiation emitted from the earth's surface, but they also radiate strongly towards it. The net radiative loss is greatly reduced and the fall in surface temperature is much smaller.

A rather different illustration is provided by radio aerials. These are often constructed so as to radiate strongly in certain directions. It follows from Kirchhoff's law that they must also absorb strongly radiation incident from that direction. Thus the polar diagrams[32] for reception and transmission must be identical.

Kirchhoff's law is often vulgarized to 'good absorbers are good emitters', but it must never be forgotton that the identity of the absorptivity and the emissivity only holds when both refer to the same wavelength and temperature. The variation with wavelength can be of great importance. Glass for instance, is transparent in the visible but is a strong absorber in the far infra-red. This is one of the reasons why it is possible for the temperature in a glass-house to be considerably higher than that outside. The short wave, high temperature radiation of the sun passes through the glass and is absorbed by the contents which become warm. Being much cooler than the sun, the greater part of their radiation is in the far infra-red, to which the glass is opaque. The glass therefore acts as an efficient radiation screen for the radiation emitted from within and so reduces the heat losses. The 'one-way' effect of the glass-house depends on the change of wavelength which is effected when the incoming energy is absorbed and re-emitted.

8.9.4. The Stefan-Boltzmann Laws

We have now discussed the properties of materials in their

[32] The plot of radiated or received power against angle.

interaction with radiation. We now return to the application of thermodynamics to the radiation itself. Since radiation exerts a pressure, we may trap it in a cylinder with perfectly reflecting walls and do work on it by compression just as we would with a normal thermodynamic fluid.

To deduce the Stefan–Boltzmann laws we perform an *isothermal* compression. This we do by introducing into the cylinder a minute speck of black material which is kept at a constant temperature T by thermal contact with an external reservoir (Fig. 8.12). If the compression is sufficiently slow, the radiation may be considered as

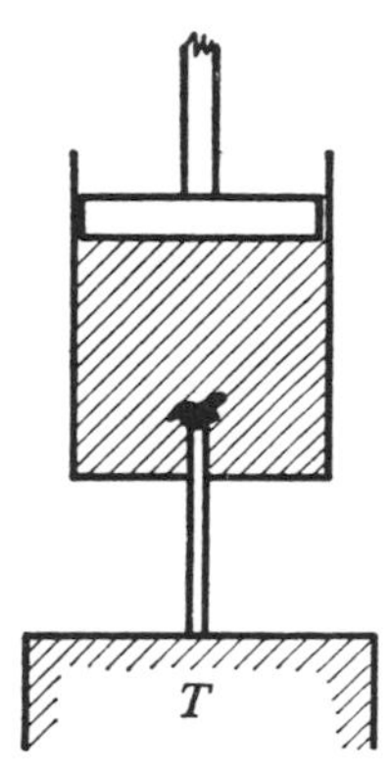

Fig. 8 . 12. Isothermal compression of black radiation.

being in thermal equilibrium with the speck at all times. It will therefore always be black and characteristic of temperature T.

The radiation has a total energy density

$$u = \int_0^\infty u_\lambda \, d\lambda$$

and exerts a pressure $p = \frac{1}{3}u$. The total internal energy of the system is

$$U = uV. \tag{8.58}$$

We know also that the quality of the radiation is a function of temperature only so that,

$$\left(\frac{\partial u}{\partial V}\right)_T = 0. \tag{8.59}$$

We apply the first law in the form

$$\mathrm{d}U = T\,\mathrm{d}S - p\,\mathrm{d}V. \tag{8.60}$$

Differentiating, we have

$$\left(\frac{\partial U}{\partial V}\right)_T = T\left(\frac{\partial S}{\partial V}\right)_T - p.$$

Substituting from (8.58) and using a Maxwell relation

$$u + V\left(\frac{\partial u}{\partial V}\right)_T = T\left(\frac{\partial p}{\partial T}\right)_V - p$$

where the second term on the left is zero by (8.59). Substituting for p,

$$u = T\,\frac{1}{3}\left(\frac{\partial u}{\partial T}\right)_V - \frac{u}{3}$$

or

$$4u = T\,\frac{\mathrm{d}u}{\mathrm{d}T}$$

where we have replaced the partial differential with an exact one since u is a function of T only. Integrating we obtain

▶ $$u = AT^4 \tag{8.61}$$

where A is a constant. (We assume that $u=0$ at $T=0$.)

Now we have shown that the spectral emissive power of a black body is related to the spectral energy density of equilibrium radiation by

$$e_\lambda = \tfrac{1}{4}cu_\lambda. \qquad \text{(Equation (8.57))}$$

Integrating over all wavelengths we obtain the total energy emitted per second per unit area by a black body

▶ $$e_b = \tfrac{1}{4}cu = \sigma T^4. \tag{8.62}$$

Equations (8.61) and (8.62) are known as the *Stefan–Boltzmann laws*, or simply as the Stefan laws. σ is the Stefan–Boltzmann constant and has the value 56·9 nW m^{-2} K^{-4}.

8.9.5. Wien's Laws

In the last section, we derived an expression for the *total* energy density of black body radiation. We now turn our attention to the

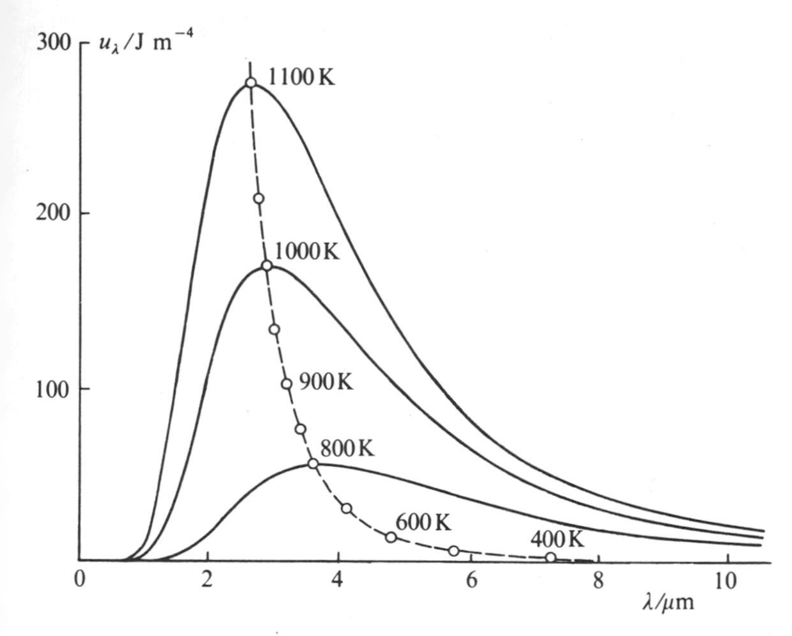

Fig. 8 . 13. The spectral energy density of equilibrium radiation as a function of wavelength for different temperatures. The broken curve shows the locus of the maxima.

way in which that energy is distributed with wavelength. Figure 8.13 shows the behaviour of u_λ which is found experimentally. Two characteristics should be noted. As the temperature is raised, (a) the energy density at any fixed wavelength always increases; and (b) the region of highest energy density moves to shorter wavelengths. The Stefan–Boltzmann law tells us that the area under the curves must be proportional to T^4.

To derive Wien's laws we perform an *adiabatic* compression of the radiation. During the isothermal compression we used in the derivation of the Stefan laws, the continued blackness of the radiation was assured by the presence of the black speck. We must first show that radiation remains black if compressed in the absence of any black body. We do this by considering the following series of processes:

Place black radiation characteristic of temperature T_1 in a cylinder with perfectly reflecting walls. Let the initial energy, entropy, and volume of the radiation be U_1, S_1, and V_1.

(*a*) Compress reversibly and adiabatically from V_1 to V_2 doing work W_a. The changes in entropy and internal energy are $\Delta S_a = 0$ and $\Delta U_a = W_a$.

(*b*) Introduce a black speck. If the radiation after compression were no longer black, an irreversible change to blackness would now take place accompanied by an increase in entropy, $\Delta S_b > 0$. Since no work is done and no heat enters the system, $\Delta U_b = 0$.

(*c*) Keeping the speck in the cylinder, expand adiabatically and reversibly back to V_1 doing work W_c. In this process we have $\Delta S_c = 0$ and $\Delta U_c = W_c$. Let the internal energy and entropy now be U_2 and S_2.

Now at all times the pressure depends only on the total energy density so that the pressure is the same function of volume during the expansion as it was during the compression. Therefore, $W_c = -W_a$, and, since no heat enters the system, $U_2 = U_1$ and $u_2 = u_1$. But the radiation is black both at the beginning and at the end of the series of processes; and, according to Stefan's law, the energy density of black radiation has one degree of freedom only (temperature). Thus, the facts that in the initial and the final states

(*a*) the volume is the same,
(*b*) the radiation is black, and
(*c*) the energy density is the same,

imply that the initial and final states are identical. Then, $S_2 = S$ and $\Delta S_b = 0$. Therefore, there could have been no irreversible change in (*b*) and the radiation must have been black after the compression.

Having proved this theorem, we are now able to dispense with our black speck during adiabatic changes knowing that the radiation will remain black. The importance of this is that with no body which can absorb and emit radiation in the container, we may consider the behaviour of individual rays or spectral components separately, knowing that there is no means of redistributing energy between them.[33]

We first find how the wavelength of a particular spectral component is changed by compression of the radiation. The change of wavelength occurs through the Doppler effect at the moving surfaces

[33] In quantum terms, this means that there is now no coupling between the various modes of the container so that we may examine how each mode is perturbed by change of volume as if no other modes were present.

of the container. This is most simply calculated by confining the radiation in a perfectly reflecting sphere of variable radius. At each reflection at the moving surface any ray suffers a change in wavelength

$$\mathrm{d}\lambda = 2\,\frac{\lambda}{c}\,\frac{\mathrm{d}r}{\mathrm{d}t}\cos\theta$$

where $\mathrm{d}r/\mathrm{d}t$ is the rate of increase of the radius, and θ is the angle between the ray and the radius at the point of reflection (Fig. 8.14). The ray suffers $c/(2r\cos\theta)$ such reflections per second, so that

$$\frac{\mathrm{d}\lambda}{\mathrm{d}t} = \frac{\lambda}{r}\,\frac{\mathrm{d}r}{\mathrm{d}t}$$

and integrating,

$$\lambda/r = \text{const.}$$

The wavelength scales in proportion to the radius.[34]

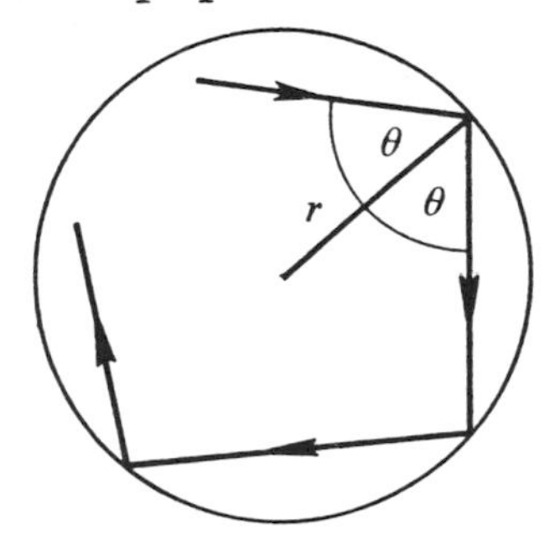

Fig. 8 . 14. The adiabatic compression of radiation in a perfectly reflecting sphere.

Still referring to the radius as our independent variable, we now find how the characteristic temperature of the radiation (as a whole) changes. Since the compression is adiabatic and reversible, the first law becomes

$$\mathrm{d}U = -p\,\mathrm{d}V,$$

or

$$\mathrm{d}(uV) = -\tfrac{1}{3}u\,\mathrm{d}V,$$

giving

$$4\,\frac{\mathrm{d}V}{V} = -3\,\frac{\mathrm{d}u}{u}.$$

[34] In the language of quantum mechanics, each eigenstate suffers an adiabatic change. No transitions occur, the wavelength of each photon scaling so as always to match the boundary conditions.

Integrating

$$V^4u^3 = \text{const.},$$

or

$$r^4u = \text{const.} \tag{8.64}$$

But, by Stefan's law,

$$u = AT^4. \tag{8.65}$$

Combining (8.63), (8.64), and (8.65),

▶ $$\lambda T = \text{const.} \tag{8.66}$$

It is very important to be clear about the significance of λ and T in (8.66). As the radiation is compressed it remains black but its characteristic temperature changes. At the same time, the wavelength *of any particular ray* changes so that the product λT remains constant. Thus λ refers to a particular spectral component of the radiation whereas T only has significance for the radiation as a whole.

We now consider the effect of compression on the spectral components between λ and $\lambda + \mathrm{d}\lambda$. The spectral density, the wavelength and the width of this band will be changed by the compression (Fig. 8.15). We apply the first law again, but this time to these spectral components only.

$$\mathrm{d}(U_\lambda \, \mathrm{d}\lambda) = -p \, \mathrm{d}V$$

$$\mathrm{d}(u_\lambda \, \mathrm{d}\lambda V) = -\tfrac{1}{3}u_\lambda \, \mathrm{d}\lambda \, \mathrm{d}V$$

$$\mathrm{d}(u_\lambda \, \mathrm{d}\lambda)V + u_\lambda \, \mathrm{d}\lambda \, \mathrm{d}V = -\tfrac{1}{3}u_\lambda \, \mathrm{d}\lambda \, \mathrm{d}V$$

$$\frac{\mathrm{d}(u_\lambda \, \mathrm{d}\lambda)}{u_\lambda \, \mathrm{d}\lambda} = -\frac{4}{3}\frac{\mathrm{d}V}{V} = -4\frac{\mathrm{d}r}{r},$$

giving

$$u_\lambda \, \mathrm{d}\lambda r^4 = \text{const.}$$

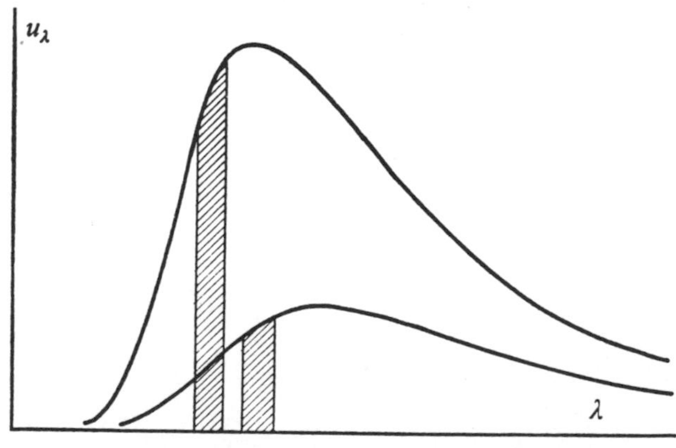

Fig. 8.15. The effect of the adiabatic compression on a band of spectral components.

But from (8.63), $\lambda \propto r$ so that $d\lambda \propto r$ also. We therefore have

$$u_\lambda r^5 = \text{const.}$$

Using (8.64) and (8.65), this gives

$$u_\lambda T^{-5} = \text{const.} \qquad (8.67)$$

We have thus shown that the group of spectral components we have isolated obeys the equations

$$\left.\begin{aligned} \lambda T &= \text{const.} \\ u_\lambda T^{-5} &= \text{const.} \end{aligned}\right\} \qquad \text{(Equations (8.66) and (8.67))}$$

Therefore, the whole distribution must obey the equation

$$u_\lambda = T^5 f(\lambda T) \qquad (8.68)$$

where $f(\lambda T)$ is an undetermined function of the *product* λT.

From this equation we see that for any given temperature the maximum of u_λ occurs where f is a maximum corresponding to a constant value of the product λT. The constant may be found experimentally or derived from quantum theory which gives an explicit expression for $u_\lambda(\lambda, T)$ in terms of fundamental constants.[35] The value found is

▶ $$\lambda_{\text{max}} T = 2{\cdot}9 \text{ mm K}. \qquad (8.69)$$

This is Wien's displacement law. Substitution of $T = 6000$ K for the optical temperature of the sun gives a maximum at 500 nm in the visible. For materials at room temperature the radiation is most intense at a wavelength of about 10 μm in the far infra-red.

We may write (8.68) in a slightly different form:

$$u_\lambda = \lambda^{-5}(\lambda T)^5 f(\lambda T),$$

or

▶ $$u_\lambda = \lambda^{-5} g(\lambda T) \qquad (8.70)$$

where $g(\lambda T)$ is another undetermined function of the *product* λT. This is *Wien's distribution law*, and gives the functional form of the spectral energy density of black body radiation.[35] To go further than this is not possible with thermodynamics. The explicit

[35] It should be noted that if we define a spectral energy density in terms of frequency u_ν such that $u_\nu \, d\nu = u_\lambda \, d\lambda$, and seek the wavelength at which u_ν is a maximum, λ'_{max}, then $\lambda'_{\text{max}} T$ is again constant, but with a slightly different value. The student should satisfy himself as to why this is so.

expression for u_λ (Planck's law), as derived from quantum theory[36] is

$$u_\lambda = \frac{c_1}{\lambda^5} \frac{1}{e^{c_2/\lambda T} - 1}$$

where c_1 and c_2 are constants. This has the form required by Wien's law.

All the results we have derived are in complete agreement with experiment. The great success with which thermodynamics is able to treat thermal radiation demonstrates how widely its laws are valid and, historically, was one of its early triumphs.

[36] See A. J. Pointon, *Introduction to Statistical Physics*, pp. 51–3, Longmans, 1967.

9. Applications to Some Irreversible Changes

Although the thermodynamic parameters of a system may only be defined when the system is in thermodynamic equilibrium, it is often possible to obtain useful results for systems which pass through non-equilibrium states provided that the initial and final states are equilibrium states and provided that sufficient information is given about the constraints applying to the irreversible changes. The Joule and Joule–Kelvin expansions of a gas are processes of this sort. In other processes, irreversibilities are involved in such a way that reversible thermodynamics cannot be applied rigorously. The thermoelectric effects are an example of this class, and we include a discussion of them here because it is instructive to examine in detail the difficulties involved.

9.1. The Joule Expansion

A Joule expansion is one in which the system exchanges no heat or work with its surroundings. That is, it takes place under the condition of complete isolation. Such is the case when a gas in a thermally isolated vessel expands into a vacuum, as illustrated in Fig. 9.1, and for this reason, a Joule expansion is also often called a 'free' expansion. To apply thermodynamics we must express the constraints which characterize the change in terms of state functions, for this will allow us to relate the initial and final states of the gas.[1] The appropriate condition follows immediately from the first law. Since $đQ = đW = 0$, we have

$$dU = 0,$$

or

$$U = \text{const.}$$

[1] For a simple fluid the equation of state defines a surface in p–V–T space. In the absence of any constraint, the whole of the surface is accessible to the system, and two variables are required to specify its state. When the constraint is applied, only one degree of freedom remains and the system can only move on a line on the surface representing the equation of state.

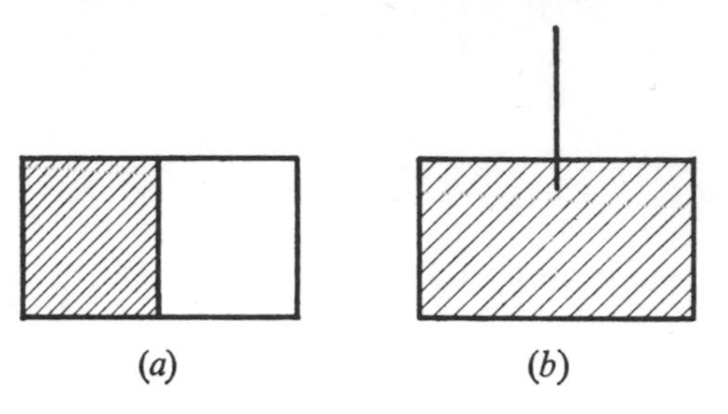

Fig. 9.1. A Joule expansion.

(*a*) A gas is kept to one part of a vessel by a partition, the other part being evacuated. (*b*) The partition is removed, and the gas expands irreversibly to fill the whole vessel.

Thus, in a Joule expansion, the system is constrained to move on a surface of constant internal energy.

We may now use this constraint to find the consequences of a Joule expansion. Suppose, for example, that we wish to calculate the temperature change produced by a Joule expansion of a gas. We would proceed as follows:

$$\left(\frac{\partial T}{\partial V}\right)_U = -\left(\frac{\partial U}{\partial V}\right)_T \left(\frac{\partial T}{\partial U}\right)_V. \tag{9.1}$$

Although the changes are irreversible, we may *always* use the first law in the form

$$\mathrm{d}U = T\,\mathrm{d}S - p\,\mathrm{d}V$$

where we may integrate from the initial to the final state by any convenient reversible path. Thus

$$\left(\frac{\partial U}{\partial T}\right)_V = C_V \tag{9.2}$$

and

$$\left(\frac{\partial U}{\partial V}\right)_T = T\left(\frac{\partial S}{\partial V}\right)_T - p$$

$$= T\left(\frac{\partial p}{\partial T}\right)_V - p. \tag{9.3}$$

Substituting (9.2) and (9.3) in (9.1),

► $$\left(\frac{\partial T}{\partial V}\right)_U = -\frac{1}{C_V}\left[T\left(\frac{\partial p}{\partial T}\right)_V - p\right] = -\frac{T^2}{C_V}\frac{\partial}{\partial T}\left(\frac{p}{T}\right)_V. \tag{9.4}$$

$(\partial T/\partial V)_U$ is known as the *Joule coefficient*. For a finite change in volume the total temperature change is found by integrating (9.4):

$$\Delta T = -\int_{V_1}^{V_2} \frac{1}{C_V}\left[T\left(\frac{\partial p}{\partial T}\right)_V - p\right] \mathrm{d}V. \tag{9.5}$$

These results are perfectly general. To evaluate the coefficient or integral in any particular case, the heat capacity has to be known and the equation of state (or equivalent information) has to be used to reduce the other terms on the right. From the mathematical point of view it is clear what is required. If we select T and V as the quantities of interest (the variables on the left of (9.4)), we must express the right-hand side in terms of the same variables so that we may (in principle) separate the variables and perform the integration. As it stands, the right-hand side contains all three variables, p, V, and T. The equation of state enables us to eliminate the one we do not want.

It should be noted that substitution of the equation of state of the perfect gas, equation (8.10), yields a Joule coefficient of zero. This must necessarily be so, since we originally defined the perfect gas as one for which U is a function of T only so that (9.1) vanishes immediately. For real gases, however, except at extremely high pressures, a Joule expansion always results in cooling. The physical reason for this may be seen as follows. In (9.1) the second term on the right is an inverse heat capacity and is necessarily positive. The first term on the right represents the change in internal energy with volume *when the temperature is kept constant*. According to the kinetic theory, when the temperature is kept constant, all contributions to the energy from kinetic terms or from degrees of freedom which are *internal* to the molecules remain unchanged. Thus, when the volume is changed, the internal energy can only alter by virtue of the change in distance between the molecules; that is, by contributions derived from the potential energy of the molecules due to their separation. This is illustrated in Fig. 9.2. At large distances there are weak attractive forces and the potential energy increases with separation. As the separation is decreased the forces eventually become repulsive and the potential energy again increases. The point of lowest potential energy, r_0, is the equilibrium separation at absolute zero, and corresponds to the density of the solid. Gases always have an average intermolecular separation greater than this, and the potential energy increases with volume, $(\partial U/\partial V)_T$ is positive, and the gas cools on expansion.

The entropy change accompanying the Joule expansion follows immediately from the first law and the condition of constant U:

$$\mathrm{d}U = T\,\mathrm{d}S - p\,\mathrm{d}V = 0$$

whence

$$\Delta S = \int_{V_1}^{V_2} \frac{p}{T}\,\mathrm{d}V. \tag{9.6}$$

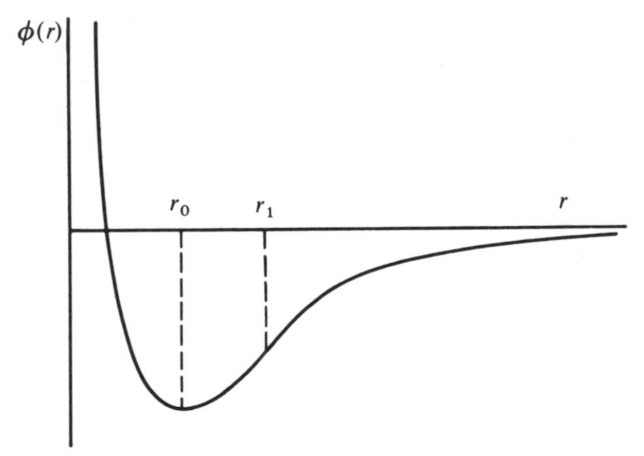

Fig. 9.2. The potential energy of two molecules as a function of their separation *r*.

For the perfect gas this becomes

$$\Delta S_m = R \ln \frac{V_2}{V_1}$$

which agrees with the general expression (8.13) since, for the perfect gas, there is no change of temperature. As $đQ = 0$, the increase in entropy is associated with the irreversibility of the expansion.

9.2. The Joule–Kelvin Expansion

The Joule–Kelvin (or Joule–Thomson) expansion is a steady flow process in which a gas is forced through a porous plug or a throttle valve under conditions of thermal isolation from the surroundings. It is represented schematically in Fig. 9.3. In passing through the plug the gas expands and the pressure drops from p_1 to p_2. The pressures on either side are kept constant, for example, by pistons moving in cylinders at the appropriate rates. Since this is a steady flow process in which no external work is done during the expansion, in which no heat is exchanged with the surroundings and in which kinetic and potential energies may normally be ignored, the constraint becomes simply that the specific enthalpy of the fluid is conserved. (See section 3.8.) The temperature drop may then be calculated as follows:

$$\left(\frac{\partial T}{\partial p}\right)_h = -\left(\frac{\partial T}{\partial h}\right)_p \left(\frac{\partial h}{\partial p}\right)_T. \qquad (9.7)$$

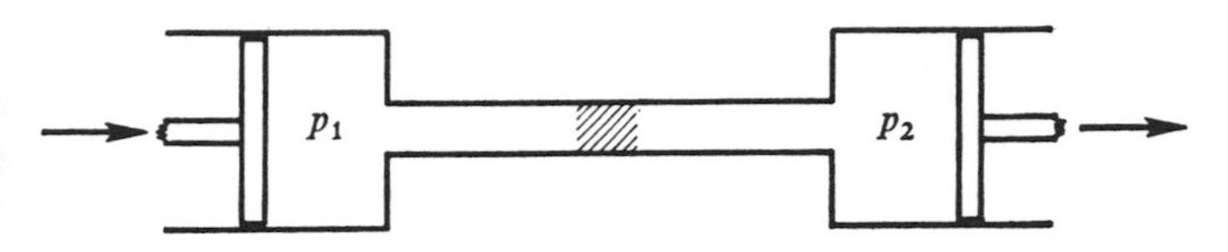

Fig. 9.3. The Joule–Kelvin expansion.

But

$$\mathrm{d}h = T\,\mathrm{d}s + v\,\mathrm{d}p \qquad (9.8)$$

so

$$\left(\frac{\partial h}{\partial T}\right)_p = T\left(\frac{\partial s}{\partial T}\right)_p = c_p \qquad (9.9)$$

and

$$\left(\frac{\partial h}{\partial p}\right)_T = T\left(\frac{\partial s}{\partial p}\right)_T + v$$

$$= -T\left(\frac{\partial v}{\partial T}\right)_p + v. \qquad (9.10)$$

Substituting (9.9) and (9.10) in (9.7) and changing from specific quantities,

▶ $$\mu = \left(\frac{\partial T}{\partial p}\right)_h = \frac{1}{C_p}\left[T\left(\frac{\partial V}{\partial T}\right)_p - V\right] = \frac{T^2}{C_p}\frac{\partial}{\partial T}\left(\frac{V}{T}\right)_p. \qquad (9.11)$$

μ is known as the *Joule–Kelvin coefficient*. For a finite change in pressure, the change in temperature is

$$\Delta T = \int_{p_1}^{p_2} \frac{1}{C_p}\left[T\left(\frac{\partial V}{\partial T}\right)_p - V\right]\mathrm{d}p. \qquad (9.12)$$

The entropy change accompanying the expansion follows directly from (9.8). Remembering that $\mathrm{d}h=0$,

$$\left(\frac{\partial s}{\partial p}\right)_h = -\frac{v}{T},$$

$$\Delta S = \int_{p_1}^{p_2} \frac{V}{T}\,\mathrm{d}p. \tag{9.13}$$

For a perfect gas the Joule–Kelvin coefficient is zero and the change of entropy

$$\Delta S_m = R \ln \frac{p_1}{p_2}$$

which agrees with (8.12) since there is no change of temperature. For the perfect gas this is entirely associated with the irreversibility of the expansion.

For real gases, a Joule–Kelvin expansion can result in heating or cooling. The physical reason for this may be seen by examining the terms in (9.7). The first term on the right is simply an inverse heat capacity and is necessarily positive. The second term may be expanded by substituting the definition of h:

$$h = u + pv$$

giving

$$\left(\frac{\partial h}{\partial p}\right)_T = \left(\frac{\partial u}{\partial p}\right)_T + \left(\frac{\partial}{\partial p}\right)_T (pv). \tag{9.14}$$

For the perfect gas both terms on the right of (9.14) would be zero. For a real gas the behaviour of the first term is similar to that of $(\partial U/\partial V)_T$ in the Joule expansion. Namely, at gaseous densities the energy increases with decreasing density, so that as the pressure drops through the porous plug the first term contributes a cooling. The second term may be of either sign, however. At low pressures the molecular volume is unimportant, and the real gas differs from the perfect gas because the intermolecular attraction reduces the pressure exerted by the gas. The attractive force is proportional to the gradient of the potential energy against separation (Fig. 9.2), so that at sufficiently low densities (corresponding to $r > r_1$) the attraction decreases with separation, $(\partial/\partial p)_T(pv)$ is negative and the last term also contributes a cooling. At high densities, on the other hand, the molecular volume becomes important and causes the product pv to increase. (This is the region for $r < r_1$.) The last term then changes sign and eventually outweighs the first so that the Joule–Kelvin expansion produces a heating. These regimes are better displayed by plotting the isenthalps of a real gas in the p–T plane (Fig. 9.4). The locus of points for which $\mu=(\partial T/\partial p)_h=0$, namely the curve which separates the region of heating from that of cooling, is called the *inversion curve*. In terms of the quantities of (9.11) this curve is determined by the condition

$$\left(\frac{\partial V}{\partial T}\right)_p=\frac{V}{T}. \tag{9.15}$$

Clearly, for a given starting temperature the greatest cooling occurs if the pressure is chosen so that the initial state lies on the inversion curve.

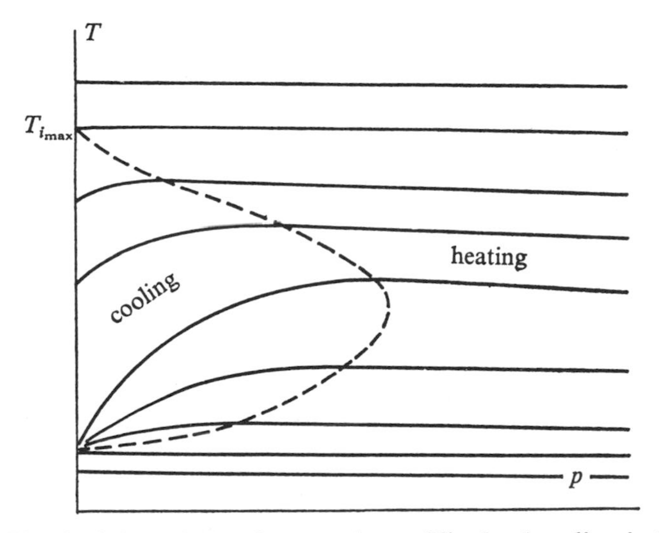

Fig. 9.4. Isenthalps for a real gas. The broken line is the inversion curve.

The condition for the inversion curve, (9.15), can, of course, be substituted into one of the approximate equations of state for a real gas to obtain its theoretical inversion curve. For example, Dietrici's equation of state in reduced units is:

$$\pi(2\phi-1)=\theta\exp\left(2-\frac{2}{\theta\phi}\right). \tag{9.16}$$

(See problem 8.13)

In reduced units, (9.15) becomes

$$\left(\frac{\partial \phi}{\partial \theta}\right)_\pi = \frac{\phi}{\theta}.$$

Applying this to (9.16) and rearranging we obtain for the inversion curve,

$$\pi = (8 - \theta) \exp\left(\frac{5}{2} - \frac{4}{\theta}\right). \tag{9.17}$$

This is compared with the experimental curve for nitrogen in Fig. 9.5. As might be expected, the agreement is not particularly good.

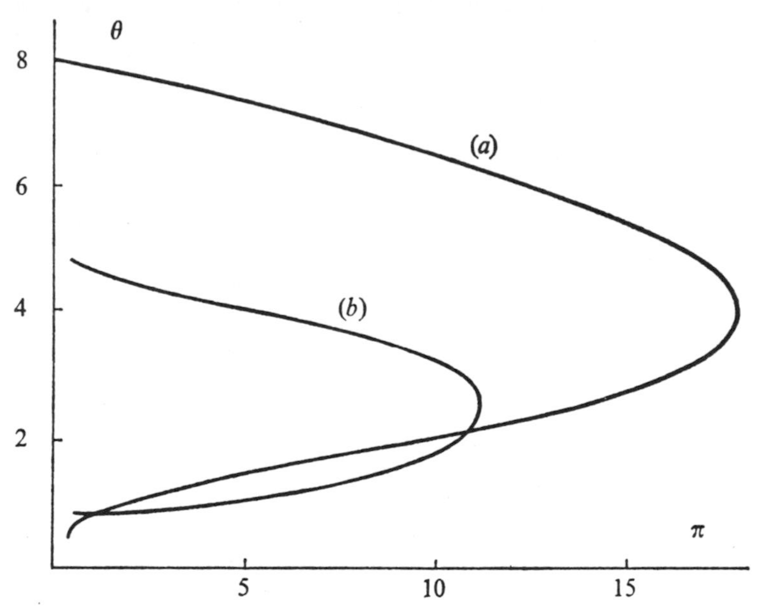

Fig. 9.5. The inversion curve of a Dieterici gas (a) compared with the experimental curve for nitrogen[2] (b).

9.3. Liquefaction of Gases

The Joule–Kelvin effect is widely used as a means of cooling in the liquefaction of gases. To obtain any cooling at all, the initial temperature must be less than the maximum inversion temperature, given by the intersection of the inversion curve with the T-axis (Fig. 9.4). Values of the maximum inversion temperatures for several gases are given in Table 9.1. Provided, then, that the initial

[2] J. R. Roebuck and H. Osterberg, *Phys. Rev.*, **48**, 450, 1935.

temperature is less than this, some cooling will be produced on expansion; but it may not be sufficient to produce any liquid. The cold, expanded gases are therefore used to cool the incoming high-pressure gas so that the expansion takes place from a lower temperature, and a lower temperature is produced. The device used for this

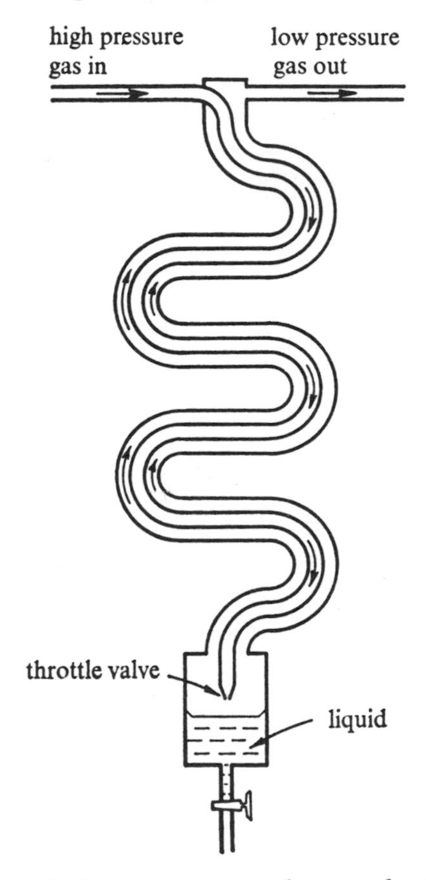

Fig. 9.6. A countercurrent heat exchanger.

is known as a *countercurrent heat exchanger* and is illustrated in Fig. 9.6. If the heat exchanger is efficient, the emerging gas will be warmed nearly to the temperature of the incoming gas. Continued operation will then cause the temperature at the throttle valve gradually to fall until liquid begins to condense out. The liquid collects in the chamber while the uncondensed gas is returned through the heat exchanger to be recompressed and recirculated.

Table 9.1 The Maximum Inversion Temperature for Several Gases*

gas	He	H_2	N_2	A	CO_2
$T_{i_{max}}$/K	23·6	195	621	723	1500

* From J. G. Kirkwood and I. Opperheim, *Chemical Thermodynamics*, p. 83, McGraw-Hill, 1961.

The efficiency y of such a simple liquefier is defined as the (mass) fraction of the incoming gas which is liquefied. In the steady state, we have a simple steady flow process in which no heat or work is exchanged with the surroundings and in which the kinetic and potential energy terms are negligible. The general equation of motion,

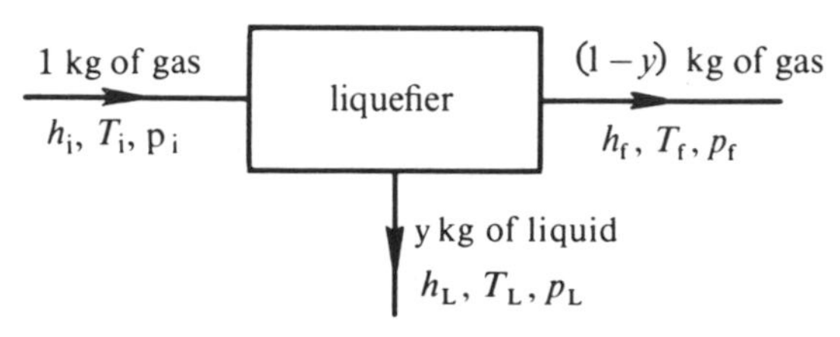

Fig. 9.7. Flow through a liquefier.

(3.31), then reduces to the simple condition that the enthalpy within the liquefier remains constant. The enthalpy of the high pressure gas entering must equal the sum of the enthalpies of the emerging gas and the emerging liquid (Fig. 9.7). If

h_i = specific enthalpy of the gas entering at T_i and p_i

h_f = specific enthalpy of the gas leaving at T_f and p_f

h_L = specific enthalpy of the liquid emerging at T_L and p_L

then

$$h_i = y\,h_L + (1-y)\,h_f$$

or

$$y = \frac{h_f - h_i}{h_f - h_L}. \tag{9.18}$$

Let us assume that we have chosen T_i and that the heat exchanger is efficient so that $T_f = T_i$. The exhaust gas would normally emerge at about atmospheric pressure so that p_f and therefore h_f are determined. p_L will only differ from p_f through any pressure drop in the heat exchanger (which would normally not be large) and is therefore also determined. Now the liquid in the chamber is in equilibrium with its vapour, so that its temperature is determined by the pressure. Hence, T_L and therefore h_L are fixed. Thus, in the expression for the efficiency, only h_i may be varied. For the maximum efficiency, h_i must be a minimum. Since T_i is fixed, the condition is

$$\left(\frac{\partial h_i}{\partial p_i}\right)_{T_i}=0.$$

But

$$\left(\frac{\partial h}{\partial p}\right)_T=-\left(\frac{\partial h}{\partial T}\right)_p\left(\frac{\partial T}{\partial p}\right)_h=-c_p\mu,$$

where μ is the Joule–Kelvin coefficient.
The point T_i, p_i must therefore be on the inversion curve. For the liquefier, the condition for maximum efficiency is the same as that for maximum cooling in a simple Joule–Kelvin expansion.

Since liquefaction using the Joule–Kelvin effect is only possible if the initial temperature is below the maximum inversion temperature, many gases have to be precooled. Usually, other liquid gases are used for this. Hydrogen ($T_{i_{max}}=195$ K), for example, is usually precooled with nitrogen (n.b.p. $=77{\cdot}3$ K). Helium ($T_{i_{max}}=23{\cdot}6$ K) is usually precooled first with nitrogen and then with hydrogen (n.b.p. $=20{\cdot}3$ K) boiling under reduced pressure at about 14 K.

As a means to gas liquefaction, the Joule–Kelvin effect has the great advantage that there are no moving components in the low temperature parts of the liquefier. As may be seen from the isenthalps of Fig. 9.4, the cooling also becomes greater as the temperature is reduced. However, when precooling is necessary, the liquefier becomes complicated and expensive to run. For this reason, some liquefiers obtain the initial cooling by causing the gas to undergo a quasi-reversible adiabatic expansion in an engine or turbine. In all gases this produces a cooling :

$$\left(\frac{\partial T}{\partial V}\right)_S=-\left(\frac{\partial T}{\partial S}\right)_V\left(\frac{\partial S}{\partial V}\right)_T=-\frac{T}{C_V}\left(\frac{\partial p}{\partial T}\right)_V. \tag{9.19}$$

(All the terms on the right are positive.) The physical reason for the cooling is simply that during the expansion the gas molecules rebound from a receding surface, the surface where the expansion occurs, so that their velocity is reduced and the temperature falls. With a heat exchanger, an expansion engine could be used to liquefy a gas directly ; but as may be seen from (9.19), the cooling decreases as the temperature falls so that the expansion engine is usually only used to obtain the precooling. The gas then passes to a Joule–Kelvin stage where a further expansion brings about liquefaction. This combination is used in the Collins helium liquefier which produces liquid helium directly from high pressure gas at room temperature.

9.4. Thermoelectricity

Thermoelectricity cannot be treated rigorously within the framework of reversible thermodynamics. We nevertheless include a discussion of it here since the difficulties involved occur in other systems also, and an examination of them is instructive in demonstrating the limitations of the theory we have developed.

9.4.1. The Thermoelectric Effects

When a temperature gradient is set up in an electrical conductor, not only does heat flow, but an electric field is also created. A qualitative picture of the origin of this field is provided by comparing the electrons in the conductor with the molecules of a gas. If the temperature in some region of a gas is raised, the increased kinetic energy of the molecules tends to cause a local increase in pressure. However, since the gas is a fluid, local pressure variations simply cause the gas to flow until the pressure is uniform, the density varying from point to point so that the product $\rho T \propto p$ is constant. In the case of the electron gas, however, the charge on the electrons prevents such a redistribution. In the undisturbed state, the charge density of the electrons exactly balances that of the positive ion cores, and the material is electrically neutral; but if the electron density is disturbed, a space charge results, and this produces an electric field in such a direction as to restore uniformity. It is as if the modulus for deformation of the electron gas were very large, the rigidity originating in the electrical forces between the electrons and the ion cores. Consequently, when there is a temperature gradient in a conductor, there is very little redistribution of charge, only enough to set up an electric field to balance the kinetic forces. For small temperature gradients, the strength of the field is proportional to the temperature gradient:

$$\blacktriangleright \qquad \mathbf{E} = P\ \nabla T$$

where P is called the *thermoelectric power* or simply the *thermopower* of the material.[3]

[3] The conventional symbol is S, but we use P here to avoid confusion with entropy.

In practice, the thermopower is not easy to measure directly because temperature gradients will normally be present in the measuring equipment also. However, the *difference* between the thermopowers of two metals is easy to measure by constructing a circuit of wires of different metals whose junctions are at different temperatures (Fig. 9.8). The e.m.f. developed in such a circuit is

$$\mathscr{E} = \int \mathbf{E} \cdot \mathrm{d}\mathbf{r} = \int P\, \nabla T \cdot \mathrm{d}\mathbf{r} = \int_{T_0}^{T_1} P_{\mathrm{B}} \mathrm{d}T + \int_{T_1}^{T_2} P_{\mathrm{A}} \mathrm{d}T + \int_{T}^{T_0} P_{\mathrm{B}} \mathrm{d}T$$

$$= \int_{T_1}^{T_2} (P_{\mathrm{A}} - P_{\mathrm{B}})\, \mathrm{d}T. \tag{9.20}$$

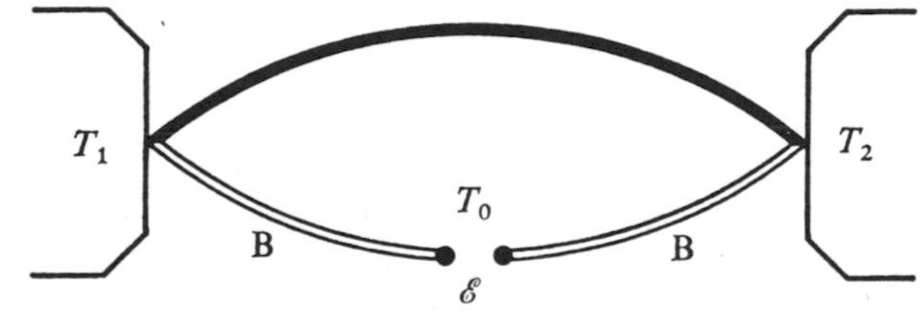

Fig. 9.8. A thermoelectric circuit of dissimilar metals with junctions at different temperatures.

This e.m.f. is known as the *Seebeck e.m.f.* Its occurrence in such a circuit is called the *Seebeck effect.* Since it is the difference of the thermopowers of two materials which is usually measured, the difference, $P_{\mathrm{A}} - P_{\mathrm{B}}$, is called the thermopower of the *couple* and is often contracted to P_{AB}.

From (9.20), it follows that if we keep the temperature of one junction fixed and vary that of the other, then

$$▶ \quad \frac{\mathrm{d}\mathscr{E}}{\mathrm{d}T} = P_{\mathrm{AB}}. \tag{9.21}$$

The Seebeck effect provides the basis of thermometry with thermocouples. The e.m.f. may usually be represented fairly well by the first few terms of a power series in the temperature difference of the two junctions. (See section 2.7.)

The *Peltier effect* is the conjugate of the Seebeck effect. Since the electrons have thermal energy (extra kinetic energy resulting from their finite temperature) current flow will necessarily be accompanied by heat flow: the electrons carry their thermal energy with them. In the Peltier effect, both the magnitude and the sign of the thermal flux depend on the details of the electronic structure of the conductor. For small current densities, the effect is linear in the current density, the constant of proportionality, Π, being called the *Peltier coefficient*.

To observe the Peltier effect we have to find some way of intercepting the heat current without disturbing the flow of electricity. Again, this is most easily done by making a junction between wires of dissimilar materials and passing an electric current across the junction. Under isothermal conditions the *difference* of the thermal currents in the two conductors will appear as heat at the junction. This will be

$$Q = I(\Pi_{\mathrm{A}} - \Pi_{\mathrm{B}}) = I\,\Pi_{\mathrm{AB}} \tag{9.22}$$

where I is the current passed and Π_{AB} is the Peltier coefficient for the couple. It should be noted that the effect is reversible so that heat may be evolved or absorbed at a junction depending on the direction of current flow.

Since the Peltier heat appears at a junction, the Peltier coefficient is often interpreted as a *voltage* (at the junction) against which work is done when charge is passed. This may sometimes be a convenient fiction, but, of course, no such voltage exists.

A third aspect of thermoelectricity is apparent as current flows along a temperature gradient: heat is generated or absorbed in the conductor in proportion to the product of the current density and the temperature gradient. The rate of heating per unit length of conductor may therefore be written

$$\dot{Q} = \mu I \frac{\mathrm{d}T}{\mathrm{d}x}. \tag{9.23}$$

This is known as *Thomson heat* and μ is the *Thomson coefficient*. Again, this effect is a consequence of the thermal energy transported by the moving charges. As the current flow carries the electrons through regions of different temperature, they have to absorb or reject heat in order to maintain thermal equilibrium with their surroundings. The Thomson coefficient is therefore simply the *heat capacity per unit charge* of the current carriers. As with the other thermoelectric coefficients, the sign and magnitude of μ depend on the details of the conductor's electronic structure.

In metals the thermoelectric effects are generally small and their usefulness has been restricted to thermometry; but in semiconductors it is possible to control the electronic structure so as to obtain large thermoelectric effects. This has made thermoelectric refrigeration practicable[4] and various devices are now on the market in which it is used. The semi-metals, As, Sb, and Bi, are intermediate to ordinary metals and the specially prepared semiconductors. Table 9.2 gives the thermoelectric coefficients of several thermocouples typical of the various classes. Thermopower is of importance in solid state physics in that it yields information about the detailed electronic structure of a material [5].

Table 9.2 Thermoelectric Coefficients for Thermocouples at 20°C

Couple	*Class*	$P_{AB}/\mu V\ K^{-1}$	Π_{AB}/mV	$\mu_{AB}/\mu V\ K^{-1}$
Cu–Ni[a]	metals	22	6	−11
Bi–Sb[a]	semi-metals	110	32	−33
p–n junction in Bi_2Te_3[b]	specially prepared semiconductors	400	120	−6

[a] *Handbook of Physics and Chemistry*, 38th edn., p. 2409, Chemical Rubber Publishing Company, 1956.

[b] H. J. Goldsmid, A. R. Sheard, and D. A. Wright, *Brit. J. Appl. Phys.* **9**, 365, 1958.

9.4.2. The Essential Irreversibility in Thermoelectricity

The reason why the thermoelectric effects cannot be treated rigorously within the framework of equilibrium thermodynamics is that, in addition to the reversible effects we have already considered, there must also always be present two irreversible processes: Joule heating due to the electrical resistance of the conductor and thermal conduction. We would only be justified in applying reversible thermodynamics if the reversible processes could be separated from the irreversible by so arranging a thermoelectric circuit that the entropy changes due to the irreversible processes could be made negligible. This cannot be done as we may show by the following argument.

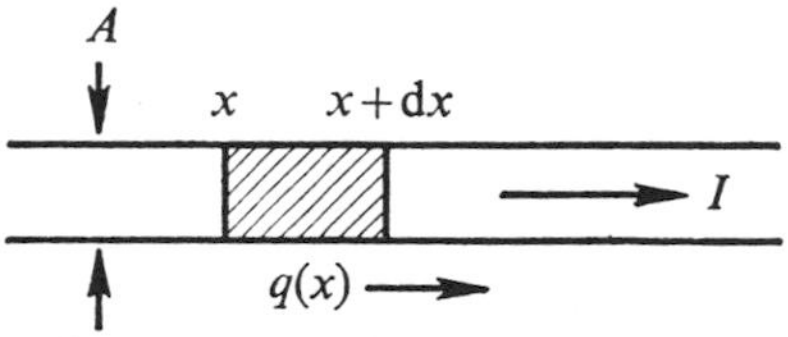

A

x $x+dx$

I

$q(x)$

Fig. 9.9. Entropy generation by an element of a wire along which there are gradients of temperature and electrical potential.

[4] See H. J. Goldsmid, *Brit. J. Appl. Phys.*, **11**, 209–17, 1960.

[5] For a discussion of the microscopic aspects of thermoelectricity, see J. M. Ziman, *Electrons and Phonons*, Oxford University Press, 1962.

Consider an element of a wire of cross-section A and length $\mathrm{d}x$ through which there flows an electric current I and along which there is a temperature gradient $\mathrm{d}T/\mathrm{d}x$ (Fig. 9.9). Let the electrical resistivity be ρ and the thermal conductivity λ. Since the element is in a *steady* state, the entropy change associated with the processes within it must appear in its surroundings.[6] This may be written in terms of the heat flow at either end:

$$\dot{S}\,\mathrm{d}x = \frac{q(x+\mathrm{d}x)}{T(x+\mathrm{d}x)} - \frac{q(x)}{T(x)}$$

where $\dot{S}$ is the rate of entropy generation per unit length of wire and q is the rate of heat flow. This may be written

$$\begin{aligned}\dot{S} &= \frac{\mathrm{d}}{\mathrm{d}x}\left(\frac{q}{T}\right)\\ &= \frac{1}{T}\frac{\mathrm{d}q}{\mathrm{d}x} - \frac{q}{T^2}\frac{\mathrm{d}T}{\mathrm{d}x}.\end{aligned} \tag{9.24}$$

But

$$q = -A\lambda\frac{\mathrm{d}T}{\mathrm{d}x}, \tag{9.25}$$

and applying the first law to the Joule and Thomson effects in the wire,

$$\frac{\mathrm{d}q}{\mathrm{d}x} = \frac{I^2\rho}{A} + I\mu\frac{\mathrm{d}T}{\mathrm{d}x}. \tag{9.26}$$

Substituting (9.25) and (9.26) in (9.24) we obtain

$$\dot{S} = \underbrace{\frac{I^2\rho}{AT}}_{\textit{Joule}} + \underbrace{\frac{I\mu}{T}\left(\frac{\mathrm{d}T}{\mathrm{d}x}\right)}_{\textit{Thomson}} + \underbrace{\frac{A\lambda}{T^2}\left(\frac{\mathrm{d}T}{\mathrm{d}x}\right)^2}_{\textit{conduction}}.$$

It is clear from the form of this that it is not possible to make the first and last terms small without making the contribution from the reversible Thomson heat small also. It is therefore impossible to

[6] When we discussed the increase of entropy associated with heat conduction in section 5.5, it was the entropy change in the *bodies* which we calculated although it was the process of conduction in the thermal resistance which caused it. We implicitly took the thermal resistance to be in a steady state, the increase in entropy associated with the irreversible processes in it being communicated, by the flow of heat, to its surroundings. All we are doing here is to generalize that argument to the case where the heat flow is not uniform.

separate the reversible and irreversible effects in thermoelectricity. Thermoelectricity is *essentially* an irreversible phenomenon and no treatment by reversible thermodynamics can be rigorous.

Lord Kelvin, who was well aware of these difficulties, obtained the thermoelectric equations by applying reversible thermodynamics to the 'reversible parts' of the thermoelectric effect while taking no account of the irreversible.[7] Others[8] have tried to take the irreversibility into account within the context of classical thermodynamics. All these approaches involve a division of the thermoelectric effect into reversible and irreversible parts and it may be shown that all contain implied assumptions which are strictly outside classical thermodynamics.[9] The more recently developed science of irreversible thermodynamics invokes microscopic ideas to provide the further information required for a rigorous treatment.[9]

Thermoelectricity is typical of the irreversible processes which have stimulated the development of irreversible thermodynamics. Its relevance to this book is that our discussion has helped to demonstrate the limitations of the classical theory. For the sake of completeness, and since no other treatment purporting to be based on reversible thermodynamics is more sound, we now give a derivation of the thermoelectric equations following that of Kelvin.

9.4.3. Kelvin's Treatment of Thermoelectricity

Consider a thermoelectric circuit of two metals, A and B, with junctions at T and $T+\mathrm{d}T$ in which the Seebeck e.m.f. is balanced by a reversible cell of e.m.f. $\mathrm{d}\mathscr{E}$ (Fig. 9.10). If we ignore the irreversible processes, we have an equilibrium situation in which charge may be passed reversibly around the circuit. For brevity we write $P_{\mathrm{AB}}=P$ and $\Pi_{\mathrm{AB}}=\Pi$.

Application of the first law[10]

Pass unit charge around the circuit. Then conservation of energy

[7] Kelvin, *Math. Phys. Papers*, **1**, 232, 1882.

[8] See, for example, R. C. Tolman and P. C. Fine, *Rev. Mod. Phys.*, **20**, 51, 1948.

[9] See S. R. de Groot and P. Mazur, *Non-equilibrium Thermodynamics*, North Holland, 1963.

[10] The irreversible parts present no difficulties here since applying the first law amounts to no more than a statement of the conservation of energy. If the irreversible terms are included, they eventually drop out and the same result is obtained.

demands that

$$d\mathscr{E} = \Pi(T+dT) - \Pi(T) + (\mu_A - \mu_B)\, dT$$

which can be arranged

$$\frac{d\mathscr{E}}{dT} = P = \frac{d\Pi}{dT} + \mu, \tag{9.27}$$

where, for brevity, we have written μ for $\mu_A - \mu_B$.

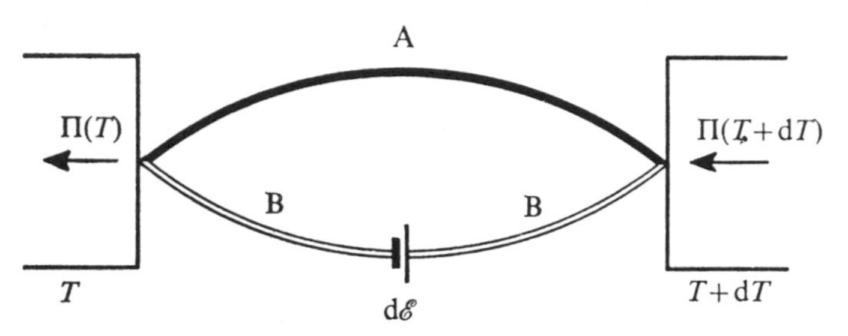

Fig. 9 . 10. Kelvin's derivation of the thermoelectric equations.

Application of the second law

Since we exclude the irreversible processes from consideration, the heat exchanges which occur when charge passes around the circuit must conserve (universal) entropy. There is no entropy change in the battery but entropy is generated in the wire by the Thomson heat. Equating the total entropy change to zero we have:

$$\underbrace{\frac{\Pi(T+dT)}{T+dT} - \frac{\Pi(T)}{T}}_{\text{From Peltier heat at the junctions}} + \underbrace{\int_T^{T+dT} \frac{\mu}{T}\, dT}_{\text{From Thomson heat in the wires}} = 0.$$

Dividing by dT,

$$\frac{d}{dT}\left(\frac{\Pi}{T}\right) + \frac{\mu}{T} = 0. \tag{9.28}$$

Eliminating the Thomson coefficients from (9.27) and (9.28),

$$P = \frac{d\Pi}{dT} - T\frac{d}{dT}\left(\frac{\Pi}{T}\right) = \frac{\Pi}{T} \tag{9.29}$$

and substituting back in (9.28),

$$\frac{dP}{dT} = -\frac{\mu}{T}. \tag{9.30}$$

(9.29) and (9.30) are the thermoelectric equations:

▶ $$P = \frac{d\mathscr{E}}{dT} = \frac{\Pi}{T}$$

▶ $$\frac{dP}{dT} = \frac{d^2\mathscr{E}}{dT^2} = -\frac{\mu}{T}.$$

The treatment by irreversible thermodynamics also leads to these equations. Their truth is well supported by experiment.

10. Change of Phase

10.1. Systems of More than one Phase

When a system consists of more than one phase, each phase may be considered as a separate system within the whole. The thermodynamic parameters of the whole system may then be constructed out of those of the component phases. If the interaction between the phases were restricted to energy exchange (flow of heat and performance of work), then the application of thermodynamics to the whole system would not lead to any essentially new results. However, if we allow new degrees of freedom within the system, such as mass transport between phases or chemical reaction between constituents, the conditions for thermodynamic equilibrium do lead to new results which are related to the restrictions which equilibrium places on the new degrees of freedom. In this chapter, we shall restrict ourselves to considering systems whose chemical composition is uniform (for example, systems of one component) but in which more than one phase is present. For simplicity, we shall again develop the general results for a system subjected to work by hydrostatic pressure only.

10.2. The General Conditions for Thermodynamic Equilibrium

Suppose that a system interacts with its surroundings. Then if heat enters, the entropy change of the system is related to the heat flow by

$$đQ \leq T_0 \, \mathrm{d}S \tag{10.1}$$

where T_0 is the temperature of the surroundings and the equality sign necessarily holds when the change is reversible. If the surroundings exert a pressure p_0 and are the only source of work, then

$$đW = -p_0 \, \mathrm{d}V. \tag{10.2}$$

Substituting in the first law,

$$\mathrm{d}U \leq T_0 \, \mathrm{d}S - p_0 \, \mathrm{d}V$$

or

$$\mathrm{d}A = \mathrm{d}U + p_0 \, \mathrm{d}V - T_0 \, \mathrm{d}S \leq 0 \tag{10.3}$$

where

$$A = U + p_0 V - T_0 S. \tag{10.4}$$

The quantity A is known as the *availability* of the system. It should be noted that it contains T_0 and p_0 which refer to the surroundings and may be quite different from the temperature and pressure of the system.

Equation (10.3) expresses the fact that in any natural change the availability of a system cannot increase. It follows that the general condition for equilibrium of a system *in given surroundings* is that the availability be a minimum. Then we must have

$$\mathrm{d}A = \mathrm{d}U + p_0 \mathrm{d}V - T_0 \,\mathrm{d}S = 0 \tag{10.5}$$

for all possible infinitesimal displacements from equilibrium. We have obtained this result directly from the law of increase of entropy as expressed in equation (10.1), by considering the interaction of a system with its surroundings.

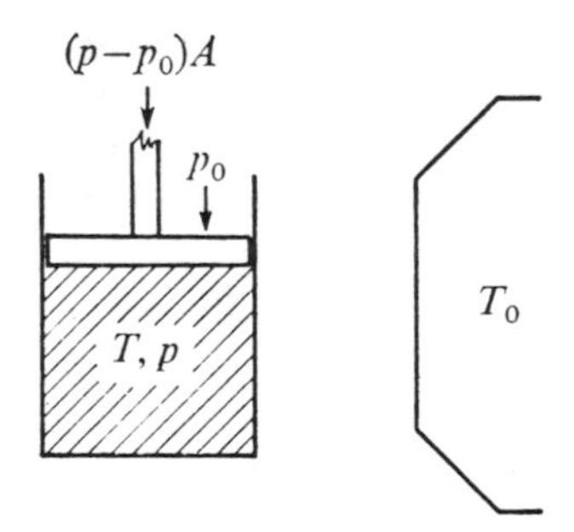

Fig. 10 . 1. Demonstration of the significance of availability.

A is known as the availability since it gives a measure of the *maximum* amount of work which may be extracted from a system *in given surroundings*. We may see this to be so by the following argument. Suppose that we place the system in a cylinder fitted with a piston so that we may subject it to a pressure p different from p_0 ; and suppose that we isolate it thermally from the surroundings so that its temperature T may also be different from T_0 (Fig. 10.1). Then, in a given change of state we shall extract the greatest possible amount of work if the change is performed reversibly. (See section 5.4.) For a small *reversible* change,

$$\mathrm{d}U = T\,\mathrm{d}S - p\,\mathrm{d}V$$

and

$$\mathrm{d}A = (T - T_0)\,\mathrm{d}S - (p - p_0)\,\mathrm{d}V. \tag{10.6}$$

Now suppose that we change the entropy of the system by operating a reversible heat engine between it and the surroundings. Then, since the process is reversible, universal entropy remains constant, and the work done by the engine is

$$đW_e = -(đQ + đQ_0) = -(T - T_0)\,dS.$$

The first term therefore represents the maximum work which is obtained from the system in the entropy change. Similarly, $(p - p_0)\,dV$ is the *net* mechanical work done on the piston. As long as $T \neq T_0$ and $p \neq p_0$ we may continue to extract work in this way and to reduce the value of A. Thus $(A - A_{min})$ is equal to the maximum amount of work which may be extracted in the given surroundings.

Now, the general condition for equilibrium, namely that the availability be a minimum, reduces to simpler forms in several important cases. We again suppose the system to be isolated from its surroundings as in Fig. 10.1, so that we may explore how A varies near equilibrium by displacing the system *reversibly* from equilibrium and using equation (10.6) to examine the consequences of the displacement. For A to be a minimum, *both* terms in (10.6) must be zero in an infinitesimal displacement. We now consider four special cases.

(a) Thermally isolated, isovolumic system

Since the system is thermally isolated, T will in general be different from T_0 so that for the first term to be zero we must have $dS = 0$. (S will, of course, be a maximum.) Since $dV = 0$, the second term is necessarily zero, and p is not directly defined.
Then (10.5) reduces to

$$dA = dU = 0.$$

Therefore, the appropriate set of conditions on the system is

▶ $$dS = 0, \qquad dV = 0, \qquad dU = 0.$$

(b) Thermally isolated, isobaric system

Again, for the first term to be zero we must have $dS = 0$. Since the volume may now change, for the second term to be zero, we require $p = p_0$, or $dp = 0$.

Then (10.5) reduces to $dA = dU + p\,dV = dH = 0$ since p is constant.

The appropriate set of conditions is

▶ $$dS = 0, \qquad dp = 0, \qquad dH = 0.$$

(*c*) *Not thermally isolated, isovolumic system*

The entropy may now change so that for the first term of (10.6) to be zero we must have $T = T_0$, or $\mathrm{d}T = 0$.

Since $\mathrm{d}V = 0$, the second term is necessarily zero and p is not directly defined.

Equation (10.5) reduces to $\mathrm{d}A = \mathrm{d}U - T\,\mathrm{d}S = \mathrm{d}F = 0$ since T is constant.

The appropriate set of conditions is

▶ $$\mathrm{d}T = 0, \qquad \mathrm{d}V = 0, \qquad \mathrm{d}F = 0.$$

(*d*) *Not thermally isolated, isobaric system*

As in (*c*), we must have $T = T_0$ and $\mathrm{d}T = 0$.

As in (*b*), since the volume may change, equilibrium requires $p = p_0$ and $\mathrm{d}p = 0$.

Equation (10.5) now reduces to

$$\mathrm{d}A = \mathrm{d}U - T\,\mathrm{d}S + p\,\mathrm{d}V = \mathrm{d}G = 0$$

since T and p are constant.

The appropriate set of conditions is

▶ $$\mathrm{d}T = 0, \qquad \mathrm{d}p = 0, \qquad \mathrm{d}G = 0.$$

The four sets of conditions for equilibrium are :

$$\left.\begin{array}{lll} \mathrm{d}S = 0, & \mathrm{d}V = 0, & \mathrm{d}U = 0 \\ \mathrm{d}S = 0, & \mathrm{d}p = 0, & \mathrm{d}H = 0 \\ \mathrm{d}T = 0, & \mathrm{d}V = 0, & \mathrm{d}F = 0 \\ \mathrm{d}T = 0, & \mathrm{d}p = 0, & \mathrm{d}G = 0. \end{array}\right\} \qquad (10.7)$$

These are entirely equivalent, and any one is sufficient to determine the equilibrium configuration. Which of these to use is entirely a matter of convenience. If a system is kept at constant temperature and pressure, the obvious choice is to minimize the Gibbs function since its accompanying conditions are automatically fulfilled.

It is important to be clear about the significance of the results we have derived. In arriving at the general condition for equilibrium, namely that the availability be a minimum, we placed no restrictions on the internal complexity of the system. We may generally expect $\mathrm{d}U$ to contain, besides T, S, p, and V, other variables related to degrees of freedom which are internal to the system. Corresponding terms do not appear in (10.6) because the system *as a whole* is subject to work by hydrostatic pressure only. Thus, the conditions for equilibrium must be thought of as placing restrictions on variables which are, at the moment, implicit in the potentials. When we come to use these results, we shall have to make them explicit.

The nature of the extrema implied by equations (10.7) are easily identified by reference to equation (10.3) and the original definitions of the potentials. For example, consider the third condition. F is defined by the equation

$$F = U - TS.$$

In general, a change in F is

$$\mathrm{d}F = \mathrm{d}U - S\,\mathrm{d}T - T\,\mathrm{d}S$$

whence,

$$\mathrm{d}A = \mathrm{d}F + S\,\mathrm{d}T + p_0\,\mathrm{d}V.$$

If T and V are given, then for A to be a minimum, F must be a minimum.

If F and T are given, then V must be a minimum.

If F and V are given, then T must be a minimum.

In this way, we may compile a list of twelve conditions for equilibrium. These are set out in Table 10.1. Again it must be emphasized that these are all equivalent. They all rest on the law of increase of entropy; but each represents the simplest way of applying the law under given conditions. It is helpful to the memory to note that the potential functions are always coupled with their proper variables, as we might expect. When the proper variables are the quantities held constant, the potential function is always a minimum, which makes it clear why they are so called from analogy with mechanical potential energy. It is also worth pointing out that if the appropriate potential is a maximum we have a situation of unstable equilibrium. We shall have an example of this in section 10.11. These conditions provide the basis for the treatment of phase change and underlie much of chemical thermodynamics.

10.3. The Condition for Equilibrium between Phases

Let us first consider a one-component system of two phases maintained at constant pressure and temperature (Fig. 10.2). This might be a liquid in contact with its vapour. If we ignore any possible surface effects at the interface, both temperature and pressure will be uniform throughout. Suppose that the masses

Table 10.1 Conditions for Stable Equilibrium

	Variables specified		*Equilibrium condition*	
often used	T	p	G	minimum
	T	V	F	minimum
	U	V	S	maximum
rarely used	S	V	U	minimum
	S	p	H	minimum
	G	T	p	maximum
	G	p	T	minimum
	F	T	V	minimum
	F	V	T	minimum
	U	S	V	minimum
	H	S	p	maximum
	H	p	S	maximum

present in each phase are m_1 and m_2 and that their specific Gibbs functions are g_1 and g_2. Then

$$G=m_1g_1+m_2g_2 \tag{10.8}$$

where G is a function of p, T, m_1, and m_2. g_1 and g_2, on the other hand, are functions of p and T only. Since p and T are constant, the condition for equilibrium reduces to

$$\mathrm{d}G=g_1\,\mathrm{d}m_1+g_2\,\mathrm{d}m_2=0. \tag{10.9}$$

As we are considering a closed system in which mass is conserved this is subject to the constraint

$$\mathrm{d}m_1+\mathrm{d}m_2=0. \tag{10.10}$$

Hence,

▶ $$g_1=g_2. \tag{10.11}$$

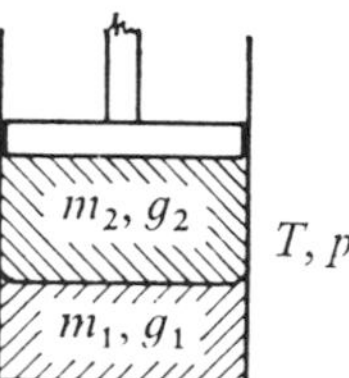

Fig. 10 . 2. A system of one-component and two phases.

This argument may readily be generalized to cases where more than two phases are present with the result that for equilibrium the specific Gibbs functions are all equal.

Although this condition was derived for a system subject to constant temperature and pressure, it holds, as we might in fact expect, whatever the external constraints. For example, if we have constant volume and temperature, the appropriate condition for equilibrium is $dF = 0$. The new constraint gives:

$$m_1 v_1 + m_2 v_2 = \text{const.}$$

or,

$$m_1\, dv_1 + m_2\, dv_2 + v_1\, dm_1 + v_2\, dm_2 = 0. \qquad (10.12)$$

The equilibrium condition gives

$$f_1\, dm_1 + f_2\, dm_2 + m_1\, df_1 + m_2\, df_2 = 0. \qquad (10.13)$$

Multiplying (10.12) by p and adding to (10.13)

$$(f_1 + pv_1)\, dm_1 + (f_2 + pv_2)\, dm_2 + m_1(df_1 + p\, dv_1) + m_2(df_2 + p\, dv_2) = 0. \qquad (10.14)$$

But in an isothermal reversible change, $df + p\, dv = 0$ so that the last two terms in parentheses are identically zero. Since we still have conservation of the total mass, (10.10) still holds and we obtain

$$(f_1 + pv_1) = (f_2 + pv_2)$$

or,

$$g_1 = g_2. \qquad (10.15)$$

A similar calculation yields the same result for any conditions of constraint for the whole system.

10.4. The Clausius–Clapeyron Equation

A simple substance normally has two degrees of freedom. If, however, we require that two phases of the substance coexist in equilibrium, then only one degree of freedom remains. The pressure and temperature of a given mass of water may be chosen at will; but if water is to be in equilibrium with its vapour, then the pressure, which is now by definition the vapour pressure, becomes a unique function of the temperature. If the pressure is increased above the vapour pressure, then the vapour will condense. If it is reduced below, the liquid will evaporate. Equation (10.11) leads immediately to an important result connecting the pressure and temperature when two phases are in equilibrium.

Consider the boundary between two phases of a substance (Fig. 10.3). At any point on the boundary the specific Gibbs functions of the two phases must be equal. In particular, this must be true at the neighbouring points a and b :

$$g_1^{(a)} = g_2^{(a)} \qquad \text{and} \qquad g_1^{(b)} = g_2^{(b)}. \tag{10.16}$$

In passing from a to b we may write,

$$\left.\begin{aligned} dg_1 &= \left(\frac{\partial g_1}{\partial p}\right)_T dp + \left(\frac{\partial g_1}{\partial T}\right)_p dT = v_1\, dp - s_1\, dT \\ \text{and} \qquad & \\ dg_2 &= \left(\frac{\partial g_2}{\partial p}\right)_T dp + \left(\frac{\partial g_2}{\partial T}\right)_p dT = v_2\, dp - s_2\, dT. \end{aligned}\right\} \tag{10.17}$$

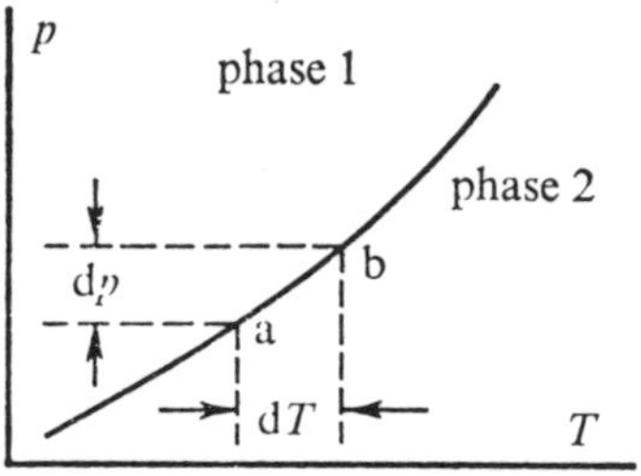

Fig. 10 . 3. Derivation of the Clausius–Clapeyron equation.

From (10.16) and (10.17)

$$v_1\, dp - s_1\, dT = v_2\, dp - s_2\, dT.$$

$$\blacktriangleright \qquad \frac{dp}{dT} = \frac{\Delta S}{\Delta V} = \frac{L}{T \Delta V} \tag{10.18}$$

where L, ΔS, and ΔV are the latent heat (absorbed), the change in entropy and the change in volume on passing from phase 1 to phase 2. Equation (10.18) is the *Clausius–Clapeyron* equation. It gives the rate at which the pressure must change with temperature for two phases to remain in equilibrium : it gives the gradient of the phase boundary in the p–T plane. It applies, to all changes of phase in which there is a discontinuity in entropy and volume at the transition. These are known as *first order* phase changes for reasons which will be explained in section 10.8. This class includes all solid–liquid, liquid–vapour, and solid–vapour transitions.

The phase diagram for a simple substance is shown in Fig. 10.4(a). The lines represent the unique relationships which must exist between the pressure and temperature if two phases are to coexist. For all three phases to coexist there can remain no freedom in the system and the condition $g_1=g_2=g_3$ leads to a unique temperature and pressure which define the *triple point*.

For most substances the gradient of the solid–liquid line is positive. The Clausius–Clapeyron equation shows that this is associated with the fact that most substances expand on melting and therefore have ΔV positive. (ΔS must, of course, always be positive because of the increase in disorder associated with melting[1].) Water is an exception in that it expands on freezing so that the solid–liquid boundary has a negative slope. Thus, in the case of water, it is possible by increasing the pressure isothermally to pass from vapour to solid to liquid (e.g. at T_1 in Fig. 10.4(*b*)) whereas, for most substances, the solid is the high-pressure phase.

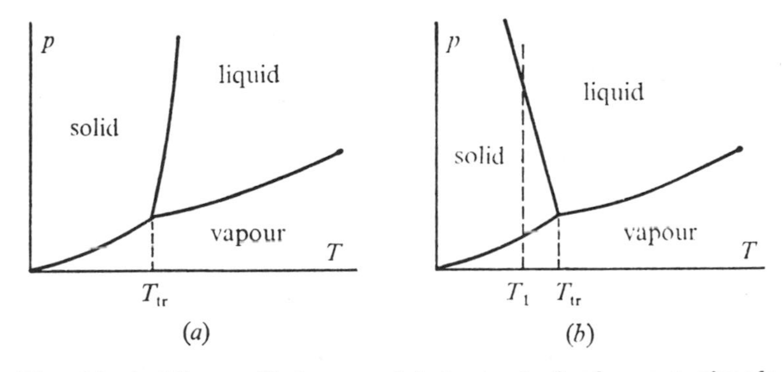

Fig. 10.4. Phase diagrams. (*a*) is typical of most simple substances, while (*b*) shows the behaviour of water which expands on freezing.

The Clausius–Clapeyron equation has been checked experimentally over a wide range of conditions in experiments on the vapour pressure of solids and liquids, and in measurements of melting curves. All the measurements have shown it to be obeyed to a high order of accuracy. Its validity provides one of the most direct experimental tests of the truth of the second law.

10.5. Integration of the Clausius–Clapeyron Equation

It is sometimes useful to have an explicit functional form which may be used as an approximation for the relationship between vapour pressure and temperature. The Clausius–Clapeyron equation itself is exact, but by making approximations which are reasonable

[1] The transition in liquid He^3 is an exception. See section 12.2.

under certain conditions it is possible to integrate it to obtain an explicit expression for the vapour pressure.

If the pressure is not too high and we are not too near to the critical point, it is reasonable to assume that the vapour obeys the perfect gas law and that the specific volume of the condensed phase is negligible in comparison with that of the vapour. With these assumptions the Clausius–Clapeyron equation becomes

$$\frac{\mathrm{d}p}{\mathrm{d}T}=\frac{Lp}{RT^2} \tag{10.19}$$

where R is the gas constant.

To proceed further we need to assume some functional form for the latent heat. There are two degrees of approximation which it is useful to make:

(*a*) The crudest approximation is to take L as constant. Over a sufficiently small temperature interval this is not unreasonable. Equation (10.19) then integrates to give

$$R\ln p=-\frac{L}{T}+A \tag{10.20}$$

where A is a constant; or

$$p=p_0\exp\left(-L/RT\right). \tag{10.21}$$

(*b*) Instead of assuming that the latent heat is constant we may make the better approximation that the specific heats of the two phases are constant. Now

$$L=T\left(S_{\mathrm{v}}-S_{\mathrm{c}}\right) \tag{10.22}$$

where the suffixes v and c indicate the vapour and the condensed phases. Differentiating *along the phase boundary* we have

$$\frac{\mathrm{d}}{\mathrm{d}T}=\left.\frac{\partial}{\partial T}\right)_p+\frac{\mathrm{d}p}{\mathrm{d}T}\left.\frac{\partial}{\partial p}\right)_T$$

whence

$$\begin{aligned}\frac{\mathrm{d}}{\mathrm{d}T}\left(\frac{L}{T}\right)&=\frac{C_{p\mathrm{v}}-C_{p\mathrm{c}}}{T}+\left[\left(\frac{\partial S_{\mathrm{v}}}{\partial p}\right)_T-\left(\frac{\partial S_{\mathrm{c}}}{\partial p}\right)_T\right]\frac{\mathrm{d}p}{\mathrm{d}T}\\&=\frac{C_{p\mathrm{v}}-C_{p\mathrm{c}}}{T}-\left[\frac{\partial(V_{\mathrm{v}}-V_{\mathrm{c}})}{\partial T}\right]_p\frac{\mathrm{d}p}{\mathrm{d}T}.\end{aligned}$$

Applying again the condition that $V_v \gg V_c$, using the perfect gas law and substituting from (10.19),

$$\frac{d}{dT}\left(\frac{L}{T}\right) = \frac{C_{pv} - C_{pc}}{T} - \frac{L}{T^2}$$

$$dL = (C_{pv} - C_{pc})\, dT. \qquad (10.23)$$

Integrating

$$L = L_0 + L_1 T. \qquad (10.24)$$

Substituting back in (10.19) and integrating again,

$$R \ln p = -\frac{L_0}{T} + L_1 \ln T + A' \qquad (10.25)$$

where A' is a constant.

It should be noted that the assumption of constant specific heats is equivalent to introducing a linear term in the temperature dependence of the latent heat, as evidenced by (10.24). For the liquid–vapour boundary, L_1 would usually be negative as $L \to 0$ at the critical temperature.

10.6. The Gibbs Functions in First Order Transitions

It is instructive to examine how the Gibbs functions of two phases behave in the neighbourhood of a transition. The specific Gibbs function for a single phase must be a continuous function of pressure and temperature. This may be represented as a *surface* in three-dimensional g–p–T space. The surfaces for two different phases will in general, intersect in a line along which the specific Gibbs functions are equal. Along this line the two phases will be in equilibrium, while away from it, the phase with the lower g value will be the stable one as is required by the condition for equilibrium. (See Table 10.1.)

If we consider the case of a simple substance which may exist in the solid, liquid, and vapour states, there will be three g surfaces which intersect in pairs to give three lines representing equilibrium between the corresponding pairs of phases. In general, there will be one point lying in all three surfaces at which all three phases are in equilibrium. The phase diagram of Fig. 10.5 is therefore a projection on the p–T plane of the lines of intersection of the g surfaces for the solid, liquid, and vapour phases. The dotted extensions through the triple point represent the continuation of the boundary between two of the phases into the region where the third becomes more stable

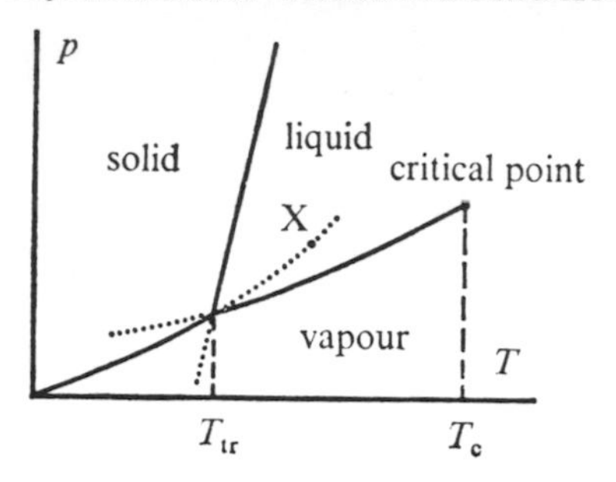

Fig. 10 . 5. A phase diagram as the projection of the intersections of the g-surfaces of the component phases.

than either of the original two. For example, the point X lies on the intersection of the solid and vapour g surfaces but does not represent a stable state because, for this value of p and T, the liquid g surface lies below the other two.

Let us now examine sections through the g surfaces. Figure 10.6 shows a section in a plane of constant T. Since, for a given p and T, the stable state is that with the lowest g, states such as Y are not stable, but may often be realized as *metastable states*. For example, if no nuclei are present to initiate condensation, a vapour may be

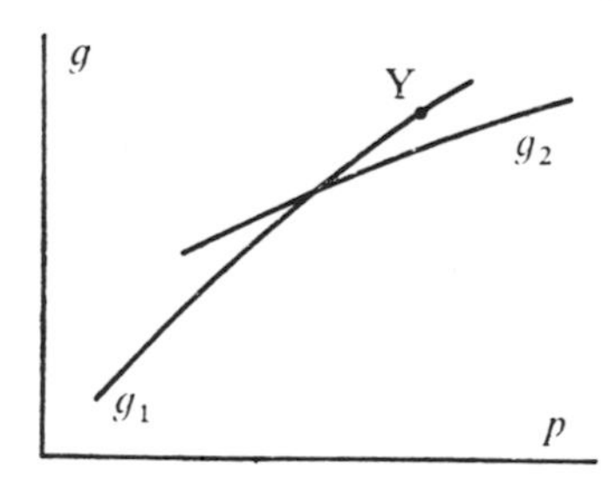

Fig. 10 . 6. A section through the g-surfaces in a plane of constant temperature.

compressed to a pressure well above the vapour pressure of the liquid without condensation taking place. It is then said to be a *supersaturated* vapour. Similarly, if a liquid is very pure it may be heated well above its boiling point without boiling taking place to produce a *superheated* liquid. The relative stability of these states is a result of surface effects which we shall discuss more fully in section 10.11; but for the moment the important point is that such states do exist, so that we are justified in extending the g functions into regions which do not correspond to a stable configuration.

Since $(\partial g/\partial p)_T = v$, the gradient of g as a function of p in a constant T section must always be positive. In the case of the solid–liquid transition, either may be the high pressure phase. It is that which has the smaller specific volume. This follows directly by the simple topological argument that since the gradients are positive, the phase with the lower g value at pressures above the intersection must have the smaller gradient. In contrast, for transitions from the vapour to the solid or liquid, the vapour must always be the low pressure phase since at a given temperature its density must always be less than those of the solid or liquid.

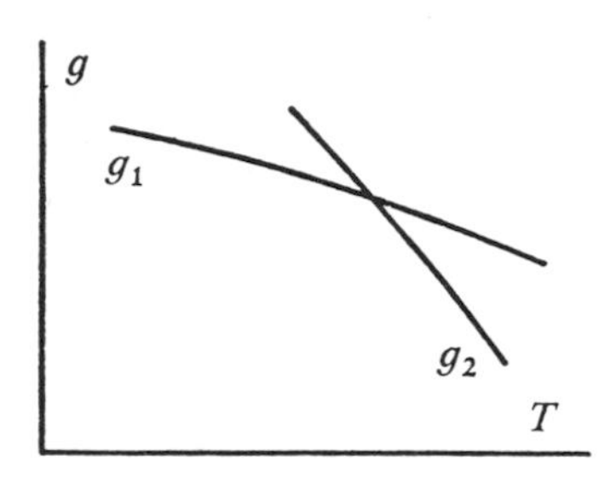

Fig. 10.7. A section through the g-surfaces in a plane of constant pressure.

A section in a plane of constant p is shown in Fig. 10.7. Because $(\partial g/\partial T)_p = -s$, the gradients are always negative. Also, since specific heats at constant pressure are always positive, the curvatures must also be negative. The topological argument now shows that the high temperature phase is always that with the greater entropy.

10.7. Critical Points

If the liquid–vapour phase change is followed in the direction of increasing temperature, it is found that the latent heat and volume change associated with the transition become smaller until they eventually vanish and it is no longer possible to identify a transition from one phase to the other. The point at which this occurs is known as the *critical point* (Fig. 10.5). Above the critical temperature, it is possible to pass continuously from the liquid to the vapour. Remembering that at a change of phase the latent heat per unit mass is

$$T\Delta s = -T\Delta\left(\frac{\partial g}{\partial T}\right)_p$$

and

$$\Delta v = \Delta\left(\frac{\partial g}{\partial p}\right)_T,$$

the absence of latent heat or volume change above the critical temperature shows that we no longer have intersecting g surfaces but that the system passes continuously along a single smooth g surface. How, then, do we change from the situation of having separate g surfaces for the liquid and vapour phases below the critical temperature to having a single g surface above it? We may gain insight into this by analysing the behaviour of a van der Waals fluid.

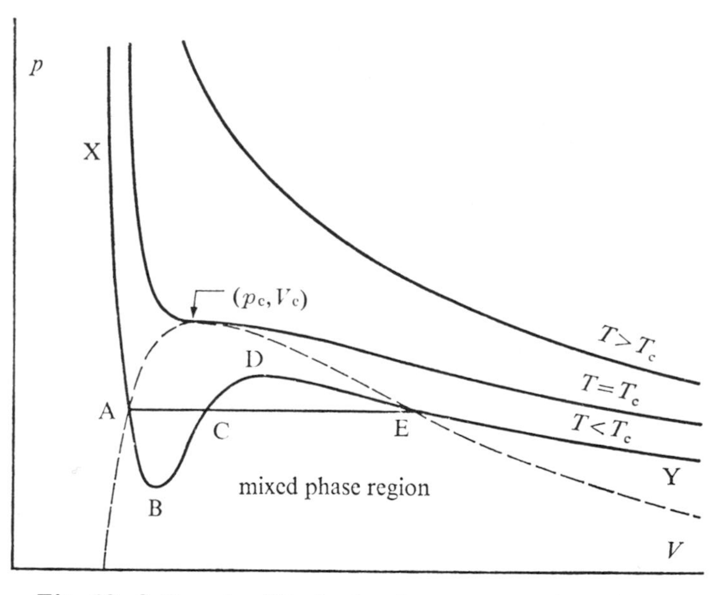

Fig. 10 . 8. Van der Waals isotherms near the critical temperature. (Not to scale.)

The form of the isotherms of a van der Waals fluid near the critical temperature is shown in Fig. 10.8. Consider the isotherm for $T < T_c$. We know that the whole of the curve YEDCBAX is not traced out by any real fluid, but that as the volume is decreased and the system passes along the curve from Y we reach some point E where the liquid begins to condense out. As the volume is further decreased, more of the substance passes into the liquid, the pressure remaining constant at the liquid vapour pressure until no more vapour remains at A. ECA is the mixed phase region. The system then follows the van der Waals isotherm along AX. In the mixed phase region we know that the specific Gibbs functions of the liquid and vapour are equal, and also that they are constant since the pressure and temperature do not change along ECA. Then, in particular, the Gibbs function for the *system* at E, where it is all vapour, and at A, where it is all liquid, must also be equal. Hence,

$$g_A = g_E. \tag{10.26}$$

Now, let us suppose that the whole of the van der Waals isotherm has physical meaning. Then we may calculate how g varies along the isotherm using

$$g(p, T) = g(p_0, T) + \int_{p_0}^{p} \left(\frac{\partial g}{\partial p}\right)_T \mathrm{d}p$$

$$= g(p_0, T) + \int_{p_0}^{p} v\mathrm{d}p \tag{10.27}$$

where we may substitute for v from van der Waals equation. The behaviour of g calculated in this way is illustrated in Fig. 10.9. The states represented by BCD in Figs. 10.8 and 10.9 are mechanically unstable, for this part of the van der Waals isotherm has a negative modulus; but the regions AB and DE may in principle be traced out as metastable states. Points B and D represent the limits of metastability for the van der Waals fluid.

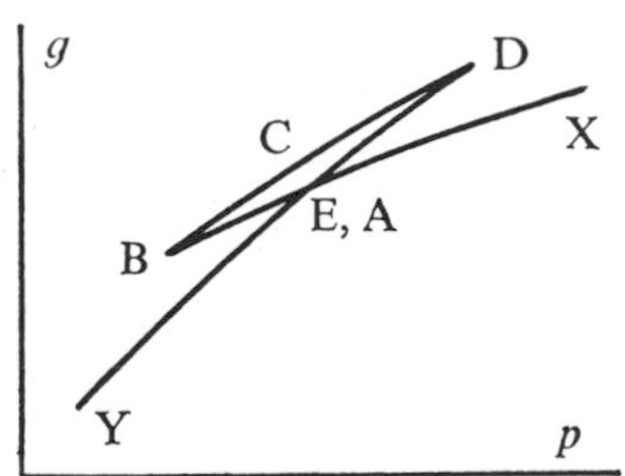

Fig. 10.9. The Gibbs function of a van der Waals fluid below the critical temperature.

Now, as the temperature is raised, the mixed phase region of the van der Waals fluid becomes smaller and eventually disappears at the critical temperature. As a consequence, the size of the closed loop in the Gibbs function and the difference in the gradients at the intersection (A, E) also become smaller and disappear at the critical point. Thus below the critical temperature the Gibbs surface of the *stable state* has a crease which becomes shallower as the critical point is approached, eventually vanishing there to give a smooth surface above it. The form of the van der Waals g surface is illustrated in Fig. 10.10.

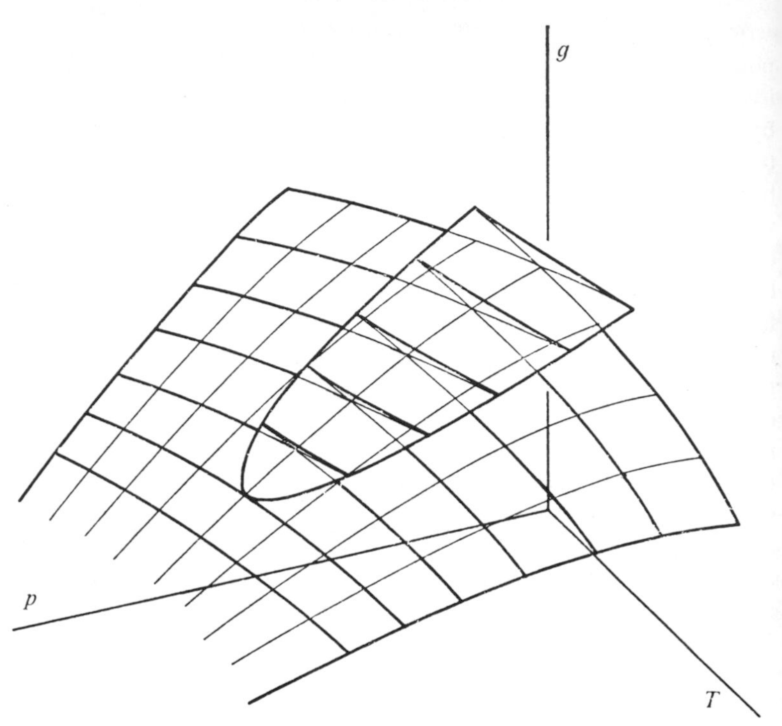

Fig. 10 . 10. The Gibbs function of a van der Waals fluid near the critical point.

It is worth noting that the vapour pressure of a van der Waals fluid may be found from (10.26), (10.27) and van der Waals equation itself. Since $g_A = g_E$, then $\int_A^E v \, dp = 0$. This means that the areas ABC and CDE in Fig. 10.8 must be equal. They may be made so by adjusting the pressure p' at which the mixed phase region occurs; but this is by definition the vapour pressure. Once the vapour pressure is found, the volume change associated with the phase change, $v_A - v_E$, and also, with the help of the Clausius–Clapeyron equation, the latent heat of the transition may be calculated. It is not surprising that experiment shows that the behaviour of these quantities in real fluids is rather different from that of the van der Waals fluid[2].

[2] See J. R. Partington, *An Advanced Treatise on Physical Chemistry*, vol. 1, section VII C, Longmans, 1949.

It is interesting to enquire whether there might be a critical point for the solid–liquid transition. Theoretical arguments would lead us to suppose that it is unlikely. If there were such a point, then by going to higher temperatures and pressures it would be possible to pass continuously from the solid to the liquid. Now the true solid has long range order: Its atoms are arranged in a definite pattern which extends throughout the solid and vests it with a particular crystal symmetry. A liquid, on the other hand, does not possess long range order. The positions of its atoms are correlated for only a few atomic spacings. It is therefore difficult to see how a continuous transition from one state to the other is possible. Unfortunately, this argument is hardly conclusive for a liquid does have short range order, and one must admit the possibility of the situation where the range of order could gradually be extended by altering the external conditions. This kind of progressive ordering certainly occurs in some systems. For example, paramagnetic salts pass from a magnetically disordered state to one of long range order without any discontinuity. On the other hand, the salts already possess a symmetry in the magnetically disordered state by virtue of their crystalline form, which makes the validity of the comparison dubious.

Experiment, however, suggests that there is no solid–liquid critical point. According to the law of corresponding states, the phase diagrams of all simple substances should be similar when plotted in reduced coordinates. The highest *effective* pressures and temperatures are therefore obtained by working with substances with as low a critical pressure and temperature as possible. The common isotope of helium, with $p_c = 2{\cdot}25$ atm and $T_c = 5{\cdot}2$ K, is an obvious choice. In one set of experiments[3] the melting curve of ^{4}He has been followed up to 7500 atm and 50 K without revealing a critical point. In another[4], the entropy difference between the solid and the liquid has been measured up to 3000 atm and 26 K and been shown to increase with temperature, whereas in approaching a critical point, it must vanish. Both these experiments suggest that it is extremely unlikely that a solid–liquid critical point exists.

10.8. Higher Order Change of Phase

The kind of phase change we have analysed in the earlier part of this chapter is characterized by discontinuous changes in the

[3] F. A. Holland, J. A. W. Huggill, and G. O. Jones, *Proc. Roy. Soc.* **A 207**, 268, 1951.

[4] J. S. Dugdale and F. E. Simon, *Proc. Roy. Soc.* **A 218**, 291, 1953.

entropy and volume at the transition. These are related to discontinuous changes in the first derivatives of the Gibbs function for the stable state of the system with respect to its proper variables p and T. Not all phase changes are of this type. It is convenient to adopt a classification scheme first introduced by Ehrenfest whereby the *order* of a transition is defined as *the order of the lowest differential of the Gibbs function which shows a discontinuity at the transition.* Table 10.2 lists the first, second, and third differentials of g and the most closely related experimental quantities in which the discontinuities appear. Figure 10.11 illustrates schematically the behaviour of the Gibbs function and its first and second derivatives in first and second order transitions.

Order of transition	g	Derivatives of g	
		First	Second
First	g, p	v, p	κ_T, p
Second	g, p	v, p	κ_T, p

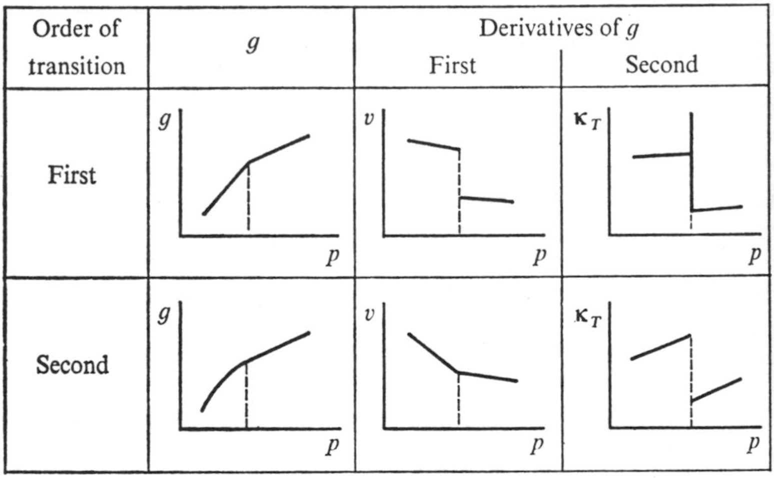

Fig. 10 . 11. The behaviour of the Gibbs function and its first two derivatives in first and second order transitions.

In section 10.4, we obtained the Clausius–Clapeyron equation for the gradient of the phase boundary by using the equality of the specific Gibbs functions for the phases in equilibrium. If we try to apply this equation to transitions of order higher than first we obtain an indeterminate result, for both numerator and denominator are zero. We may, however, obtain analogous equations for second order transitions by using the equality of the entropies or volumes at the transition. We proceed as follows:

Expanding $\mathrm{d}v$ in terms of p and T (the two variables we are interested in),

$$\mathrm{d}v = \left(\frac{\partial v}{\partial T}\right)_p \mathrm{d}T + \left(\frac{\partial v}{\partial p}\right)_T \mathrm{d}p. \tag{10.28}$$

Using $\mathrm{d}v_1 = \mathrm{d}v_2$ for an infinitesimal change along the boundary, we obtain

$$\frac{\mathrm{d}p}{\mathrm{d}T} = -\frac{\left(\frac{\partial v_2}{\partial T}\right)_p - \left(\frac{\partial v_1}{\partial T}\right)_p}{\left(\frac{\partial v_2}{\partial p}\right)_T - \left(\frac{\partial v_1}{\partial p}\right)_T} = \frac{\beta_2 - \beta_1}{\kappa_2 - \kappa_1}. \tag{10.29}$$

Expanding $\mathrm{d}s$ instead of $\mathrm{d}v$ the result is

$$\frac{\mathrm{d}p}{\mathrm{d}T} = \frac{1}{vT}\frac{c_{p2} - c_{p1}}{\beta_2 - \beta_1}. \tag{10.30}$$

These equations:

▶ $$\frac{\mathrm{d}p}{\mathrm{d}T} = \frac{\Delta\beta}{\Delta\kappa} = \frac{1}{vT}\frac{\Delta c_p}{\Delta\beta} \tag{10.31}$$

are known as *Ehrenfest's equations*. It is worth noting that, in effect we have used the usual procedure for expressions which are indeterminate because numerator and denominator tend to zero; namely, we have replaced the numerator and denominator by their first differentials, in the first case with respect to p and in the second, T.

Table 10.2 First, Second, and Third Order Transitions

The table lists, for each order of transition, the differential coefficients of g and the most closely related experimental quantities in which discontinuity appears.

	Discontinuity appears in:				
Order	*Differentials of g*		*Corresponding experimental quantities*		
First	s	v	s	v	
Second	$\left(\frac{\partial s}{\partial T}\right)_p$	$\left(\frac{\partial v}{\partial T}\right)_p$	c_p	β	κ
	$\left(\frac{\partial s}{\partial p}\right)_T$	$\left(\frac{\partial v}{\partial p}\right)_T$			
Third	$\left(\frac{\partial^2 s}{\partial T^2}\right)_p$	$\left(\frac{\partial^2 v}{\partial T^2}\right)_p$	$\left(\frac{\partial c_p}{\partial T}\right)_p$	$\left(\frac{\partial \beta}{\partial T}\right)_p$	$\left(\frac{\partial \kappa}{\partial T}\right)_p$
	$\frac{\partial^2 s}{\partial p\,\partial T}$	$\frac{\partial^2 v}{\partial p\,\partial T}$	$\left(\frac{\partial c_p}{\partial p}\right)_T$	$\left(\frac{\partial \beta}{\partial p}\right)_T$	$\left(\frac{\partial \kappa}{\partial p}\right)_T$
	$\left(\frac{\partial^2 s}{\partial p^2}\right)_T$	$\left(\frac{\partial^2 v}{\partial p^2}\right)_T$			

10.9. Some Examples of Higher Order Phase Changes

Unfortunately, very few systems showing higher order transitions approach the idealized behaviour illustrated in Fig. 10.11. Usually the gradient of the specific heat becomes infinite on one or both sides of the transition, and it is often difficult to decide to which of the idealized classes a particular system best belongs. Some examples of transitions of various orders are given in the following list.

First order

Solid–liquid, solid–vapour, and liquid–vapour phase changes (sections 10.4–7).
The superconducting transition in a magnetic field (section 10.9 3).
Some allotropic transitions in solids (e.g., iron, Fig. 10.12).

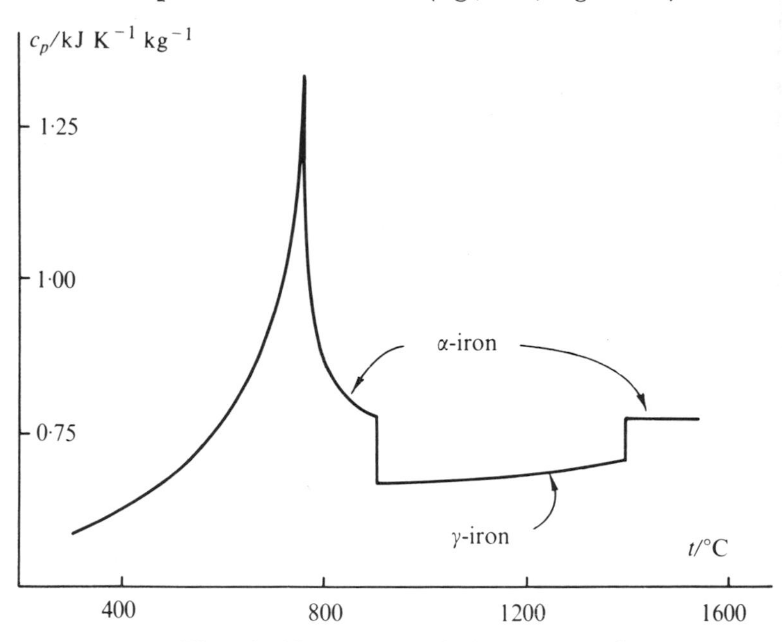

Fig. 10 . 12. The specific heat of iron.[5]

Second order

The superconducting transition in zero field (section 10.9.3).
The superfluid transition in liquid helium (section 10.9.2).
The order-disorder transition in β-brass (Fig. 5.4).

[5] From results quoted in J. B. Austin, *Indust. Eng. Chem.*, **24**, 1225, 1932.

Third order

The Curie point of many ferromagnets (e.g., iron, Fig. 10.12).

We shall now discuss some of these in more detail.

10.9.1. Phase Changes in Solid Iron

Iron is interesting in that it shows both first and third order transitions in the solid (Fig. 10.12). The third order transition is the change from the magnetically ordered ferromagnetic state to the disordered paramagnetic state, the total area under the specific heat anomaly being related to the entropy change associated with the magnetic ordering. (See section 5.6.) At higher temperatures, the two first order transitions are associated with changes in crystalline structure. Below 906°C and above 1400°C the α phase is stable, while between these temperatures the γ phase is stable. The g surfaces for the α and γ phases therefore intersect at these temperatures while the third order transition corresponds to a discontinuous change in the curvature of the g surface at the Curie point.

10.9.2. The Superfluid Transition in Liquid Helium

Figure 10.13 shows the phase diagram for the common isotope of helium ^{4}He. Unlike all other elements, ^{4}He and the lighter isotope

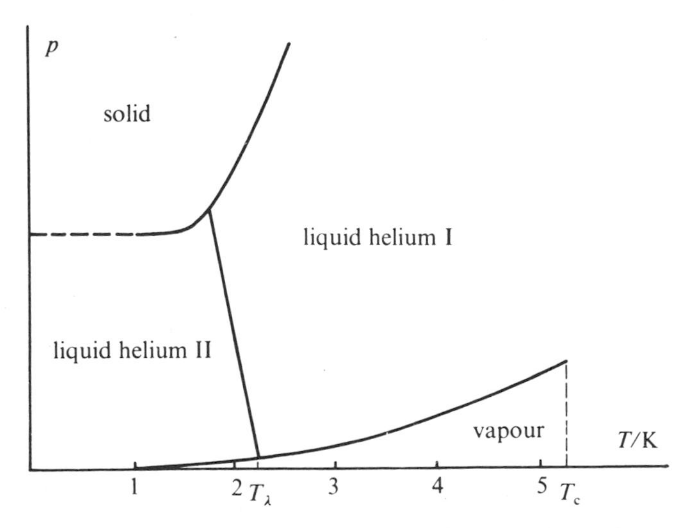

Fig. 10.13. The phase diagram for the common isotope of helium, ^{4}He. (Not to scale.)

^{3}He remain liquid to absolute zero. The reason for this is straightforward. To become solid, the atoms must be confined to proper sites on a crystal lattice. This will involve restricting their linear motion within some length Δx, which will be of the order of the atomic spacing. But this restriction may only be achieved by giving the atoms momentum Δp of a magnitude given approximately by the uncertainty principle[6]: $\Delta p\, \Delta x \sim \hbar$. This momentum corresponds to a *zero point energy* $E = (\Delta p)^2/2m$. This will be larger for helium which is a small, light atom, than for elements higher in the periodic table. Helium is also an inert gas with a closed outer shell of electrons so that the interatomic forces are very weak and the energy available for restricting the atoms to their proper positions for the solid is correspondingly small. In the case of helium, the zero point energy is greater than the energy available for bringing about solidification, and unless the effect of the interatomic forces is enhanced by applying a large pressure, the helium remains liquid to absolute zero. In hydrogen, the interatomic forces are much larger, while the heavier inert gases have a smaller zero point energy by virtue of their greater mass. The uncertainty principle also explains why the vapour pressure of the lighter isotope is higher than that of the heavier. Their respective normal boiling points are: ^{3}He, 3·19 K; ^{4}He, 4·21 K.

Unlike the lighter isotope, however, ^{4}He has two liquid phases known as helium I and helium II (Fig. 10.13). The former, the high temperature phase, is in all respects a normal liquid. Helium II, however has an extremely high thermal conductivity and, in some respects, behaves as if it has an extremely small viscosity, for which reasons it has been called 'superfluid'[7]. The transition to the superfluid state is well defined; but there is no change in density nor can any latent heat be detected. The specific heat, however, shows a strong anomaly (Fig. 10.14). It rises rapidly below the transition and apparently falls discontinuously at it.

When a specific heat anomaly has a shape like that for the superfluid transition in helium or that of the third order transition in iron, the phase change is known, on account of the shape of the specific heat curve, as a λ *transition*, and the temperature at which it occurs as the λ-*point*.

[6] D. Bohm, *Quantum Theory*, Chapter 5, Prentice Hall, 1960.

[7] Helium II also has other extraordinary properties and has been much studied. See: J. Wilks, *The Properties of Liquid and Solid Helium*, Oxford University Press, 1967.

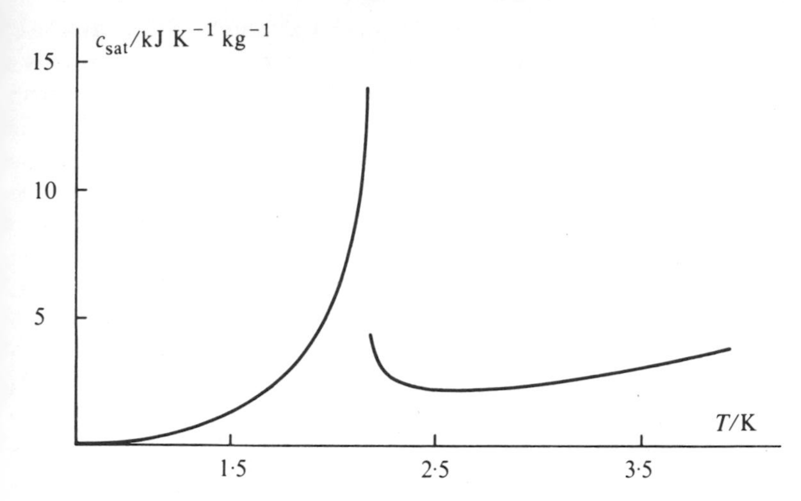

Fig. 10.14. The specific heat of liquid helium under its vapour pressure.[8]

10.9.3. The Superconducting Phase Change

We shall give a fairly detailed analysis of the superconducting phase change both because of its intrinsic interest and also because it provides the only example of an ideal second order transition.

Many metals, when cooled to a sufficiently low temperature, become superconducting[9]. In the superconducting state they are characterized by two properties:

(*a*) Zero electrical resistance,
(*b*) Perfect diamagnetism[10]. (Complete exclusion of magnetic flux.)

The superconducting state may be destroyed and the normal state restored by

(*a*) Raising the temperature,
(*b*) Applying a magnetic field greater than some critical value,

or a combination of both (*a*) and (*b*).

[8] H. C. Kramers, J. D. Wasscher and C. J. Gorter, *Physica*, **18**, 329 1952; and R. W. Hill and O. V. Lounasmaa, *Phil. Mag.*, **2**, 143, 1957.

[9] For a detailed discussion of the phenomenon of superconductivity, see A. C. Rose-Innes and E. H. Rhoderick, *Introduction to Superconductivity*, Pergamon, 1969.

[10] We are describing here the *basic* properties of superconductors. In *type II superconductors* these basic properties are modified so that a magnetic field can penetrate and perfect conductivity persist. See reference 9.

Perfect diamagnetism is not implied by infinite electrical conductivity. It is true that a changing magnetic field induces eddy currents in the surface of a conductor and that these act to screen the field changes from the interior ; but even in a perfect conductor, the surface currents decay and the field changes eventually penetrate[11]. In contrast, with a superconductor, surface currents induced by a magnetic field persist indefinitely so long as the material remains superconducting, and, what is more remarkable, if the superconductivity is destroyed by raising the field above the critical value and then restored by reducing the field, the surface currents *reappear* and all the flux is *expelled* from the interior.

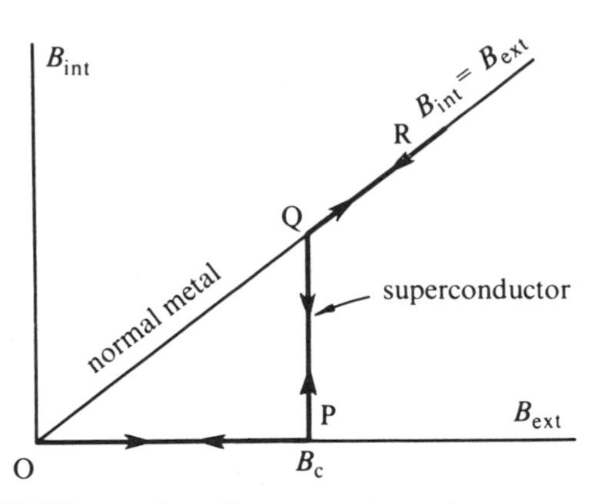

Fig. 10 . 15. The perfect diamagnetism of a superconductor.

The perfect diamagnetism of a superconductor is illustrated in Fig. 10.15. As the external field is raised, induced surface currents screen the field changes from the interior and the magnetic induction in the material remains strictly zero until the critical field is reached at P. Superconductivity is then destroyed and flux enters until $B_{int} = B_c$ at Q. For $B_{ext} > B_c$, $B_{int} = B_{ext}$ and the system moves along QR as it would for a normal metal. If the field is reduced, the surface currents reappear with the restoration of superconductivity at $B_{ext} = B_c$, the flux is expelled and the system returns along QPO. Of the two basic characteristics of superconductivity, perfect conduction and perfect diamagnetism, the latter is, in fact, the more fundamental.

[11] See A. B. Pippard, pp. 64–6 of 'The Dynamics of Conduction Electrons', in *Low Temperature Physics*, edited by C. De Witt, B. Dreyfus, and P. G. de Gennes, Gordon and Breach, 1962.

The sharpness of the transition to the superconducting state is strongly affected by the presence in the superconductor of strains and impurities. There is no doubt, however, that in the absence of such extraneous effects it is extremely sharp and reversible. For example, it has been shown by direct experiment that the transition in tin in zero magnetic field ($T_c = 3{\cdot}73$ K) takes place reversibly over a temperature interval smaller than 10^{-4} K.

Strictly, we must consider the superconductor to be a system of three degrees of freedom; p, V; T, S; $\mathbf{B}$, $\mathbf{m}$. In the thermodynamic analysis below we shall retain all of these; but superconductivity is not strongly affected by hydrostatic pressure and it is often possible to disregard the first two variables. We shall simplify the mathematics by dropping the vector notation for the magnetic quantities.

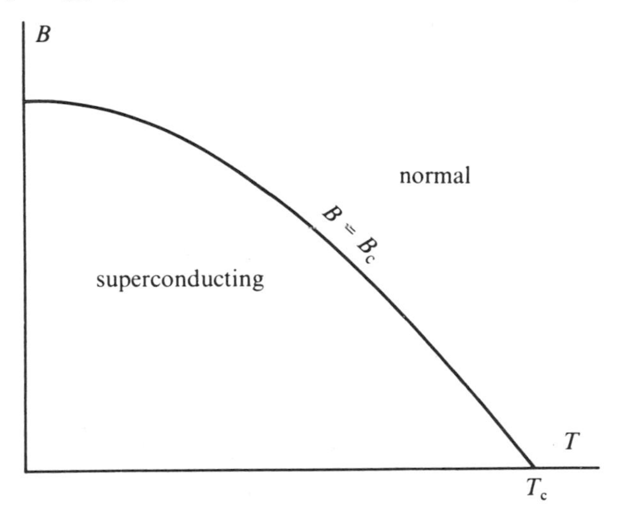

Fig. 10 . 16. The phase diagram of a superconductor.

Figure 10.16 shows the phase diagram for a superconductor in the points on the boundary between the superconducting and B–T plane. (Here, and in what follows, we use B for B_{ext}.) At all normal phases the transition is first order except at $B = 0$ and $T = 0$. (The latter, as we shall see in chapter 12, is required by the third law.) The value of the magnetic field which, at a given temperature, destroys superconductivity is known as the *critical field* B_c. In contrast, the *critical temperature* T_c is normally taken as the transition temperature in *zero* field. We now apply the thermodynamics which we have developed for change of phase.

It is convenient to choose T, B, and p as independent variables. The appropriate potential function is

$$g = u - Ts + pv - Bm \tag{10.32}$$

where m is now the magnetic moment per unit mass, and u is defined so that $\mathrm{d}u = T\,\mathrm{d}s - p\,\mathrm{d}v + B\,\mathrm{d}m$. (See section 3.5.4.)
Then,

$$\mathrm{d}g = -s\,\mathrm{d}T + v\,\mathrm{d}p - m\,\mathrm{d}B. \tag{10.33}$$

At equilibrium, $\mathrm{d}G = 0$, (since $\mathrm{d}T = \mathrm{d}p = \mathrm{d}B = 0$) which gives the condition that on the phase boundary

$$g_\mathrm{s} = g_\mathrm{n} \tag{10.34}$$

where s and n refer to the superconducting and normal phases respectively.

Before discussing properties on the phase boundary, it is worth digressing for a moment to derive a simple expression for the difference of the Gibbs functions of a superconductor in the normal and superconducting states.
From (10.33),

$$g(B) = g(0) - \int_0^B m\,\mathrm{d}B.$$

For the superconductor, as long as it remains superconducting, we may put

$$m = -Bv/\mu_0$$

since this gives $B_\mathrm{int} = 0$ and therefore corresponds to perfect diamagnetism.
Therefore

$$g_\mathrm{s}(B) = g_\mathrm{s}(0) + vB^2/2\mu_0 \tag{10.35}$$

for $B \leqslant B_\mathrm{c}$, and at the transition, using (10.34),

$$g_\mathrm{n}(B_\mathrm{c}) = g_\mathrm{s}(0) + vB_\mathrm{c}^2/2\mu_0. \tag{10.36}$$

But the normal metal has a negligible susceptibility so that to a very good approximation

$$g_\mathrm{n}(B) = g_\mathrm{n}(0)$$

and

► $$g_\mathrm{n}(0) - g_\mathrm{s}(0) = vB_\mathrm{c}^2/2\mu_0. \tag{10.37}$$

Returning now to consider the phase boundary, we may derive from (10.33) three analogues of the Clausius–Clapeyron equation taking the independent variables in pairs and applying the condition (10.34) as we did for a two-parameter system. (Section 10.4.) This gives:

$$p, T \quad (B \text{ constant}) \quad \left(\frac{\partial p}{\partial T}\right)_B = \frac{\Delta s}{\Delta v} = \frac{s_n - s_s}{v_n - v_s} \tag{10.38}$$

$$B, T \; (p \text{ constant}) \quad \left(\frac{\partial B_c}{\partial T}\right)_p = -\frac{\Delta s}{\Delta m} = -\mu_0 (s_n - s_s)/v_s \, B_c \tag{10.39}$$

$$B, p \; (T \text{ constant}) \quad \left(\frac{\partial B_c}{\partial p}\right)_T = \frac{\Delta v}{\Delta m} = \mu_0 (v_n - v_s)/v_s \, B_c. \tag{10.40}$$

Of these, the first is of identical form to the Clausius–Clapeyron equation for a system with two degrees of freedom and subject to work by hydrostatic pressure. The values of s_n, s_s, v_n, and v_s which are appropriate to these equations should, of course, be those at the transition; that is, with $B = B_c$. However, these quantities are virtually independent of field, as may be seen by examining two of the Maxwell relations generated from (10.33):

$$\left(\frac{\partial s}{\partial B}\right)_{T,p} = \left(\frac{\partial m}{\partial T}\right)_{B,p} \tag{10.41}$$

$$\left(\frac{\partial v}{\partial B}\right)_{T,p} = -\left(\frac{\partial m}{\partial p}\right)_{B,T}. \tag{10.42}$$

In the absence of ferromagnetism (and ferromagnets do not become superconducting), a normal metal is only very weakly magnetic, so that m_n is essentially zero and s_n and v_n are essentially field independent. In the superconducting state we have $m_s = -v_s B_{ext}/\mu_0$ which is almost independent of temperature and pressure if the field is constant. Hence s_s and v_s are essentially field independent. It is therefore sufficient to take for these quantities, their values in zero field.

Rearranging (10.39) we see that the change in entropy at the transition is

$$s_n - s_s = -\frac{v_s B_c}{\mu_0} \left(\frac{\partial B_c}{\partial T}\right)_p = -\frac{v_s}{2\mu_0} \left(\frac{\partial (B_c)^2}{\partial T}\right)_p. \tag{10.43}$$

This vanishes at T_c where $B_c \to 0$ with a finite slope and at $T = 0$ where $(\partial B_c/\partial T)_p = 0$ (Fig. 10.16), the latter behaviour being required by the third law (chapter 12). Except in these limits, the transition is therefore first order. Differentiating (10.43) we obtain the difference of the specific heats

$$c_{ps}-c_{pn}=\frac{Tv_s}{2\mu_0}\frac{\partial^2}{\partial T^2}(B_c^2)_p. \tag{10.44}$$

From the typical temperature dependence of the critical field (Fig. 10.16) we see that equations (10.43) and (10.44) yield entropy and specific heat differences of the form shown in Fig. 10.17. Experiment shows that these equations are obeyed well by real superconductors[12].

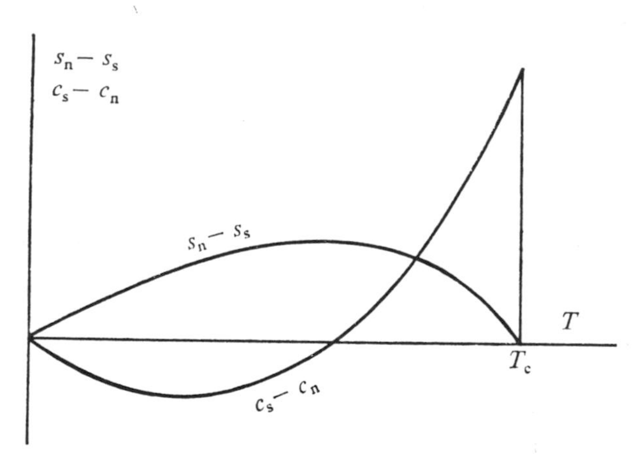

Fig. 10.17. The entropy and specific heat differences of the normal and superconducting phases.

At $T=T_c$ where the latent heat becomes zero the transition becomes second order. Evaluating the limiting forms of (10.38) by taking the differentials of the numerator and denominator with respect to T and p we obtain respectively,

$$\left(\frac{\partial T_c}{\partial p}\right)_{B=0}=vT_c\frac{\beta_n-\beta_s}{c_n-c_s}=\frac{\kappa_n-\kappa_s}{\beta_n-\beta_s} \tag{10.45}$$

which are, as we might expect, simply the Ehrenfest equations. Since the superconducting transition in zero field is of the ideal second order form (the specific heat has a simple discontinuity), it would be of great interest to verify these equations in this case.

[12] See, for example, W. S. Corak and C. B. Satterthwaite, *Phys. Rev.* **102**, 662, 1956, and C. A. Swenson, *Solid State Physics*, **11**, 41, **1960.**

Unfortunately, this test does not seem to have been made, partly because recent interest has been more concerned with a microscopic interpretation of changes at the transition and also because the changes in the expansion coefficients and compressibilities are so small as to make the experiments difficult to do with sufficient precision. For tin, for example, it is found that

$$\left(\frac{\partial T_c}{\partial p}\right)_{B=0} = -5\times 10^{-5}\ \mathrm{K\ atm^{-1}}$$

and

$$\Delta c_p = 9\times 10^{-2}\ \mathrm{J\ K^{-1}\ kg^{-1}}$$

which imply a change in the expansivity of about $5\times 10^{-8}\ \mathrm{K^{-1}}$ and a fractional change in the compressibility of about 10^{-5}.

10.10. Interpretation of Second Order Transitions

In a first order transition, the gradients of the Gibbs function for the stable state of the system change discontinuously at the transition as the system passes from the g surface of one phase to that of another at their intersection. Now, in a second order transition, the system cannot pass from one g surface to another as we may show by the following argument:

Suppose that the system did pass from one g surface to another.

(*a*) The absence of latent heat or volume change would require the gradients to be identical at the transition.

(*b*) The discontinuous change in the second order coefficients would require one curvature to be greater than the other.

(*c*) But then the surfaces would touch *but not cross* at the transition (Fig. 10.18). The same phase would always be the more stable and there would be no transition[13].

We are therefore forced to the conclusion that a single g surface is involved in a second order transition and that some property of the system causes the second derivatives to change discontinuously. This conclusion is supported by the absence in second and higher order transitions of metastable states. Superheating and supercooling have never been found in transitions other than first order, so that only there is there any evidence of the continuity of the Gibbs function for a particular phase through the transition.

[13] It is impossible for two continuous, smooth surfaces with different curvatures to intersect tangentially and cross.

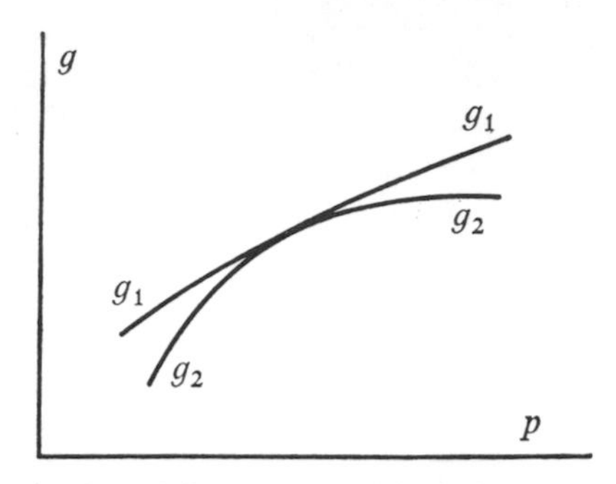

Fig. 10 . 18. A suggested but untenable behaviour for the Gibbs function in a second order change of phase.

A simple system showing a second order transition was devised by Gorter. It consists of a vessel containing a liquid and its vapour (Fig. 10.19). The walls of the vessel are slightly extensible so that its volume depends on the pressure difference between the inside and the surroundings. Also, the amount of liquid is small enough to ensure that when all the material is in the vapour state, the pressure is well below the critical pressure. Suppose, now, that the temperature is gradually raised so that the equilibrium state has an increasing amount of the substance in the vapour phase. When the last of

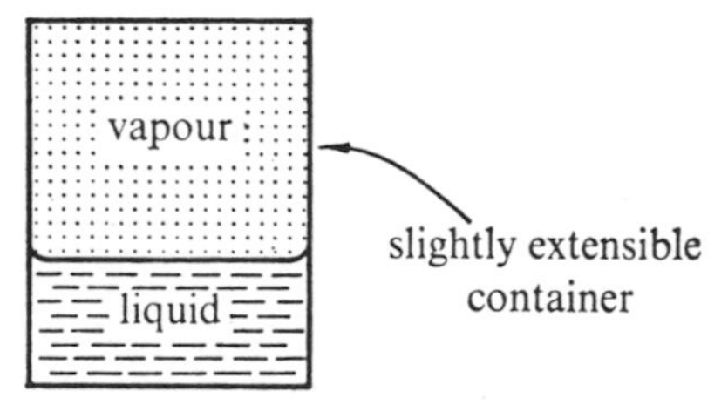

Fig. 10 . 19. The Gorter model of a second order phase change.

the liquid evaporates there will be no discontinuity in internal pressure or entropy of the system so there will be no volume change or latent heat. However, while liquid is still present, there will be a contribution to the heat capacity of the whole system from the latent heat required to vaporize the liquid. At the transition, when the last of the liquid evaporates, this contribution will suddenly vanish so that the thermal capacity of the system changes discontinuously. Similarly, below the transition, the internal pressure is simply the vapour pressure of the liquid and is independent of volume; while above the transition it follows the gas law for the vapour which will have a different temperature dependence and will also depend on volume. Therefore, the expansion coefficient and modulus of the system also change discontinuously. Thus Gorter's simple model exhibits a second order phase change. A thermodynamic analysis[14] shows it to obey the Ehrenfest equations.

Gorter's model also serves to bring out an essential feature of higher order transitions. In a first order phase change, there is a latent heat because there is a discontinuous change in the order of the system. In the second order phase change, the transition occurs when an ordering process *starts* to take place. In Gorter's model, the more highly ordered liquid phase *starts* to condense out at the transition. The theoretical treatments of liquid helium and of superconductivity have both invoked, with considerable success, *two fluid models*[15]. A *normal* fluid only is present above the transition temperature, while, below it the normal fluid gradually condenses into a *superfluid*, the condensation being complete by absolute zero. At intermediate temperatures the systems behave as if they contain a mixture of the two.

The reasons why such different kinds of behaviour are present in the higher order transition are only to be found in microscopic theory. They are connected with the different ways in which the ordering processes set in and, in particular, with the range (spatial extent) of order which the condensation requires. In superconductors the range of order is very long and this is why they show the ideal second order behaviour. In other cases (such as the ferro-magnetic transition), it is perhaps more helpful to think of the phase change as being a first order transition spread out by the presence of thermal fluctuations: As the temperature is reduced, small ordered regions first appear as transient local effects, these regions gradually become larger until, when ordering is almost complete, we have only small transient regions of disorder, and these eventually disappear as the temperature is reduced still further.

10.11. Surface Effects

So far, in our discussion of equilibrium between phases, we have not considered surface effects. As an example of a situation in which

[14] A. B. Pippard, *Classical Thermodynamics*, pp. 147–50, Cambridge University Press, 1966.

[15] For helium see, J. Wilks, *The Properties of Liquid and Solid Helium*, Oxford University Press, 1967. For superconductivity see, D. Shoenberg, *Superconductivity*, Cambridge University Press, 1960.

they are important, we shall examine the equilibrium between a liquid and its vapour when there is an interfacial surface tension.

Consider a spherical drop of radius r immersed in its vapour at temperature T_0. Let the pressure of the vapour be held constant at p_0. Then the condition for equilibrium of the system is that its total Gibbs function G be a minimum. We construct G as follows:

$$U = m_l u_l + m_v u_v + U_s \tag{10.46}$$

where the suffixes refer to liquid, vapour, and surface.

$$T_0 S = T_0(m_l s_l + m_v s_v + S_s), \tag{10.47}$$

$$p_0 V = p_0(m_l v_l + m_v v_v), \tag{10.48}$$

$$G = m_l(u_l + p_0 v_l - T_0 s_l) + m_v(u_v + p_0 v_v - T_0 s_v) + U_s - T_0 S_s. \tag{10.49}$$

Then,

$$dG = dm_l(u_l + p_0 v_l - T_0 s_l) + dm_v(u_v + p_0 v_v - T_0 s_v) + \gamma\, dA. \tag{10.50}$$

We have assumed the liquid to be incompressible, for else the first bracket would also change as a result of the change in pressure due to surface tension. But

$$dA = \frac{2v_l}{r}\, dm_l.$$

Hence

$$dG = dm_l(u_l + v_l(p_0 + 2\gamma/r) - T_0 s_l) + dm_v(u_v + p_0 v_v - T_0 s_v). \tag{10.51}$$

However, $2\gamma/r$ is the pressure difference across the surface due to surface tension, so that

$$p_l = p_0 + 2\gamma/r.$$

Then the first bracket in equation (10.51) is simply the specific Gibbs function of the liquid *evaluated at the pressure within the drop.* Thus

$$dG = g_l(p_l, T_0)\, dm_l + g_v(p_0, T_0)\, dm_v. \tag{10.52}$$

Conservation of mass requires $dm_l + dm_v = 0$, so the condition for equilibrium becomes

▶ $$g_l(p_l, T) = g_v(p_v, T). \tag{10.53}$$

That is, for equilibrium we again have the condition that the specific Gibbs functions should be equal; but g_l is evaluated at the pressure within the drop which is now different from the pressure of the vapour. Equality of specific Gibbs functions is, in fact, a general condition for equilibrium between phases. In chapter 11 we shall derive it by more general arguments.

We may use (10.53) to find the effect of surface tension on vapour pressure. If we change the radius of the drop, maintenance of equilibrium requires

$$\left(\frac{\partial g_l}{\partial p_l}\right)_T \mathrm{d}p_l = \left(\frac{\partial g_\mathrm{v}}{\partial p_\mathrm{v}}\right)_T \mathrm{d}p_\mathrm{v}$$

i.e.,

$$v_l\left(\mathrm{d}p_\mathrm{v} - \frac{2\gamma}{r^2}\,\mathrm{d}r\right) = v_\mathrm{v}\,\mathrm{d}p_\mathrm{v}$$

$$(v_\mathrm{v} - v_l)\,\mathrm{d}p_\mathrm{v} = -\frac{2\gamma}{r^2}\,v_l\,\mathrm{d}r. \tag{10.54}$$

We now make the approximations that $v_\mathrm{v} \gg v_l$, which will normally be the case, and that the liquid is incompressible. Then (10.54) becomes

$$v_\mathrm{v}\,\mathrm{d}p_\mathrm{v} = -\frac{2\gamma}{r^2}\,v_l\,\mathrm{d}r.$$

Substituting from the perfect gas law for the vapour

$$\frac{RT}{M_r}\frac{\mathrm{d}p_\mathrm{v}}{p_\mathrm{v}} = -\frac{2\gamma}{r^2}\,v_l\,\mathrm{d}r$$

where M_r is the relative molecular mass of the vapour. Integrating,

$$\frac{RT}{M_r}\ln\left(\frac{p}{p_0}\right) = \frac{2\gamma}{r}\,v_l$$

▶ $$p = p_0 \exp\frac{2\gamma v_l M_r}{rRT} \tag{10.55}$$

where p_0 is the vapour pressure over a plane liquid surface ($r = \infty$). If the change in vapour pressure is small, the exponential may be expanded to give

$$\frac{\Delta p}{p_0} = \frac{2\gamma M_r}{r\rho_l RT} \tag{10.56}$$

where ρ_l is the density of the liquid.

Generally, the effect of surface tension on vapour pressure only becomes important when the liquid surface has a small radius of curvature. For water, for example, at room temperature, $\Delta p/p_0 \simeq 1/(r/\mathrm{nm})$, so that a 100 nm radius drop has its vapour pressure increased by only 1 per cent. However, the modification of the vapour pressure becomes a crucial factor in processes involving the nucleation of one phase within another. This is because the equilibrium between a drop and its vapour is generally unstable as we may see from the form of the Gibbs function. Rearranging (10.50), with r as an independent variable, we have

$$dG = 8\pi\gamma r\,dr - [g_v(p_0, T_0) - g_l(p_0, T_0)]\,4\pi r^2 \rho_l\,dr. \qquad (10.57)$$

If, at some value of the radius, a drop would be in equilibrium with the vapour, then $\Delta g = g_v(p_0, T_0) - g_l(p_0, T_0)$ must be a *positive* constant. Then, integrating (10.57), we find that G must be of the form

$$G(r) = G(0) + 4\pi\gamma r^2 - \tfrac{4}{3}\pi\rho_l \Delta g r^3, \qquad (10.58)$$

which is illustrated in Fig. 10.20. The equilibrium condition we have derived, (10.53), therefore corresponds to a *maximum* in the free energy, which implies that the equilibrium is unstable. Drops smaller than the equilibrium radius would evaporate and those larger would grow. Raising the pressure moves the maximum to smaller radii, but the initial region of positive slope still remains. On this basis, a liquid would never condense from a supersaturated vapour; but this analysis does not, of course, take into account the presence of fluctuations in the vapour. It is possible to analyse the statistics of the motion of the gas molecules so as to work out the probability that sufficient molecules would collide simultaneously to form a droplet larger than the critical size. Such a droplet, once formed, would then grow. However, even at quite high supersaturations the probability of such an event is extremely small. For example, a small volume of water vapour at room temperature raised to 2·7 times the saturated vapour pressure would condense spontaneously in about 10^{30} years.[16] This is why clean, pure, supercooled vapours are comparatively stable.

In reality, the condensation of a supercooled vapour is almost always controlled by the presence of nuclei on which condensation can take place. In the Wilson cloud chamber, ions act as nuclei, and the droplets formed on them mark out the paths of ionizing particles. In the atmosphere, the formation of cloud takes place at a supersaturation of only a few per cent. Again, this is due to the presence in the air of condensation nuclei which range from small chemical complexes to inert particles several microns in size. The

[16] See, J. Frenkel, *Kinetic Theory of Liquids*, Oxford University Press, p. 397, 1946.

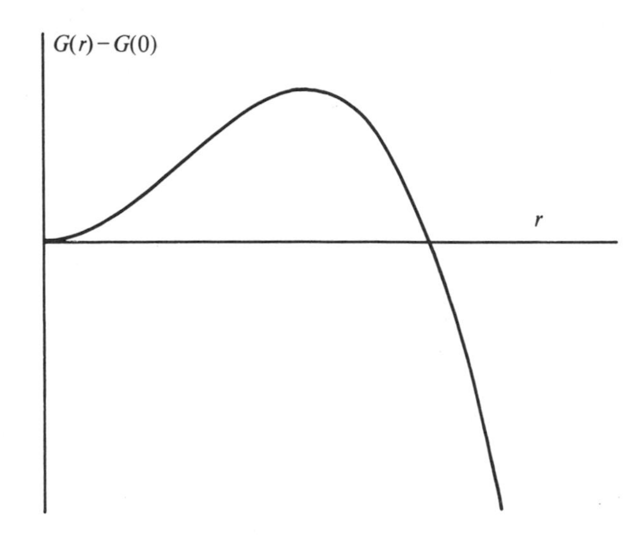

Fig. 10.20. The Gibbs function of a vapour containing a small drop of its liquid as a function of the drop radius.

variation of vapour pressure with radius is also responsible for one of the processes by which large droplets are formed in a cloud once condensation has occurred: the large droplets keep the partial pressure of the water vapour in the air down so that the smaller droplets, with a higher vapour pressure, evaporate, the net effect being to transfer water from the smaller to the larger droplets.

Similar considerations apply to the formation of the vapour phase in a superheated liquid. If a small bubble is present in a liquid the pressure within the bubble will be much greater than that of the surrounding liquid. For a bubble of vapour to be in equilibrium with its liquid, the latter must therefore be at a temperature such that its vapour pressure is much greater than would be required for equilibrium over a plane surface. Once again the equilibrium is unstable and boiling is usually promoted by the presence of nuclei.

11. Systems of Several Components

To give a proper account of systems of several components would lead us further into the field of chemical thermodynamics than is relevant to this book. However, it is important to understand how the principles that we have developed may be generalized so as to make their treatment possible. In this chapter, therefore, we give a brief account of the application of thermodynamics to systems of more than one component, and we illustrate the basic ideas and theory with a few simple applications. For the greater part we shall restrict the discussion to situations in which there is no chemical reaction; but to illustrate the principles involved, we shall include a short section on reactions in ideal gas mixtures. Those who would like a more detailed treatment of multi-component systems should pursue their interests in other texts[1].

11.1. Mixtures of Ideal Gases

According to elementary kinetic theory, the ideal behaviour is shown by a gas whose molecules have negligible size and exert negligible forces one on another. (See section 8.3.) It follows that in a mixture of ideal unreacting gases, the molecules of the different species will move independently and the properties of the mixture will be a simple combination of those which each gas would have if present alone. For example, the total pressure will be the sum of the pressures which each gas would exert individually:

$$p=\frac{RT}{V}\sum_{i} n_i \tag{11.1}$$

where V is the total volume and n_i is the number of moles of the ith gas present. The quantity

$$p_i=n_i\frac{RT}{V} \tag{11.2}$$

[1] For example, A. H. Wilson, *Thermodynamics and Statistical Mechanics*, chapters 11–13, Cambridge University Press, 1957.

is known as the *partial pressure* of the ith gas. Clearly,

$$p = \sum p_i \tag{11.3}$$

which is known as *Dalton's law*, a result which was first obtained empirically. For real gases, it holds to the same approximation as their behaviour approaches the ideal.

It also follows from the independence of the motions of the molecules of the different gases that the entropies and the thermodynamic potentials must also be additive. However, these conclusions may be demonstrated by an argument which avoids the use of microscopic ideas.

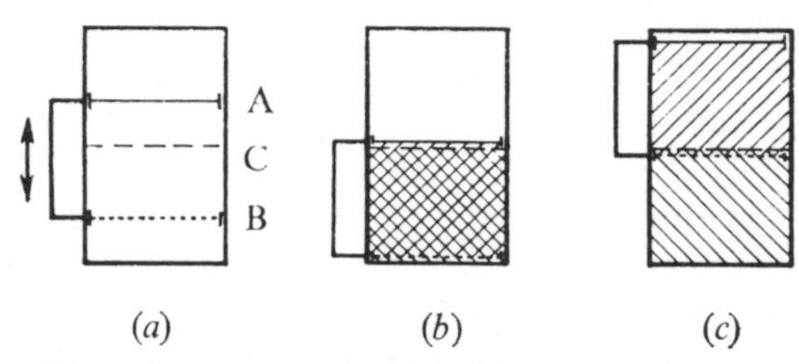

Fig. 11 . 1. Reversible mixing of two gases.

Consider a cylindrical vessel of volume $2V$ fitted with three partitions, A, B, and C (Fig. 11.1(a)). Partition C is fixed across the centre. A and B are tight fitting but moveable and are attached together so as to enclose a volume V. Suppose we have a mixture of two ideal gases, 1 and 2. Then we choose A to be impermeable, B to be permeable to 2 but not to 1 and C to be permeable to 1 and not to 2. The device now allows us to mix or separate the two gases reversibly. We start with the mixture in the lower half of the vessel, the upper half being evacuated and the coupled partitions, A and B, in their lowest position (Fig. 11.1(b)). Gas 2 passes freely through B. Then assuming the truth of Dalton's law[2], its pressure on either side of B will be the same and it will exert no force on it. Similarly, gas 1 passes freely through C but is contained between A and B and so exerts equal and opposite forces on them. There is therefore no net force on the coupled partitions. Now, if A and B are raised, gas 2 remains in the lower half of the cylinder while gas 1 is swept through C into the upper half (Fig. 11.1(c)). This is clearly a reversible process. Since the volume of the whole system is constant,

[2] This is where the assumption that the gases are perfect and unreacting is introduced. We invoke Dalton's law as an empirical fact. See problem 11.1.

the only work that *could* be done on it would be of the form $X\,\mathrm{d}x$, where X and x are the net force on and displacement of the moveable partitions. This would give $\mathrm{d}U = T\,\mathrm{d}S + X\,\mathrm{d}x$, whence we have a Maxwell relation,

$$\left(\frac{\partial S}{\partial x}\right)_T = \left(\frac{\partial X}{\partial T}\right)_x.$$

But $X = 0$ always, so that

$$\left(\frac{\partial X}{\partial T}\right)_x = 0$$

in particular, and the entropy must remain constant if the gases are separated isothermally. Further, since

$$\left(\frac{\partial T}{\partial x}\right)_S = -\frac{T}{C_x}\left(\frac{\partial S}{\partial x}\right)_T,$$

it follows that the temperature would not change even if the process were performed adiabatically. Thus, in either case, we have $\Delta S = \Delta T = 0$. Also, since no work is done, the total internal energy is unchanged, and the other potentials also. The argument is readily generalized to more than two gases. Therefore, we have the general result that the *pressure, entropy, and thermodynamic potentials of a mixture of perfect, unreacting gases are all sums of the corresponding quantities which each gas would have if present alone.*

It follows that the natural variables to choose for describing a particular component in a mixture of gases are T, S_i, V, and its *partial* pressure p_i, for then its contribution to any particular thermodynamic property of the whole system takes the form which would apply in the absence of the other components. For example, the total entropy of a mixture of perfect gases is

$$S = \sum_i S_i(p_i, T) \tag{11.4}$$

where the S_i are identical with the expressions which hold for pure, perfect gases, namely

$$S_{mi} = S_{m0i} - R \ln p_i + C_{mp} \ln T. \tag{11.5}$$

However, it is often useful in treating mixtures of gases to refer the entropy or Gibbs function of a particular component to the *total* pressure rather than to the partial pressure. This may be done by using the *molar concentrations* of the components, c_i, defined by

$$c_i = \frac{n_i}{\sum n_i} \tag{11.6}$$

where the n_i are the numbers of moles of each component present in the mixture. Then

$$\frac{p_i}{p}=\frac{n_i}{\sum n_i}=c_i, \tag{11.7}$$

which gives the molar entropy of one component in terms of the total pressure and its molar concentration as

$$S_{mi}(T, p, c_i)=S_{m0i}+C_{mp}\ln T-R\ln p-R\ln c_i \tag{11.8}$$

or

$$S_{mi}(T, p, c_i)=S_{m0i}(T, p)-R\ln c_i. \tag{11.9}$$

This form has the advantage that we have separated the expression for the entropy into two terms, one of which is invariant in systems subject to constant temperature and *total* pressure, and the other of which depends only on concentration. This makes it much easier to discuss the effects of changing the composition of a mixture.

A similar separation may be made for the Gibbs potential. For a perfect gas,

$$\left(\frac{\partial G_m}{\partial p}\right)_T=V_m=\frac{RT}{p}.$$

Integrating,

$$G_m(T, p)=G_{m0}(T)+RT\ln p \tag{11.10}$$

where G_0 is a function of T only. Then, for the components of a mixture of perfect gases we have the various useful forms for the molar Gibbs function:

$$G_{mi}(T, p_i)=G_{m0i}(T)+RT\ln p_i \tag{11.11}$$

$$G_{mi}(T, p, c_i)=G_{m0i}(T)+RT\ln p+RT\ln c_i \tag{11.12}$$

$$G_{mi}(T, p, c_i)=G_{m0i}(T, p)+RT\ln c_i. \tag{11.13}$$

11.2. The Increase of Entropy in Diffusion

Consider a vessel divided by a partition into two regions of volume V_1 and V_2 which contain two different gases, 1 and 2, both at temperature T and pressure p (Fig. 11.2). If the partition is removed there will be mechanical and thermal equilibrium between the two gases, but interdiffusion will take place so that after some time the vessel will contain a uniform mixture. This is an irreversible process and we would expect it to be accompanied by an increase in entropy[3]. Now, for perfect gases, the process of interdiffusion is

[3] The difference between this mixing process and the one we considered in section 11.1 is that here the initial and final volumes occupied by each gas are different. The gases could, of course, be recompressed reversibly into their original volumes with semipermeable membranes but this would require work.

equivalent to making each gas perform a Joule expansion to a volume (V_1+V_2), and then mixing them reversibly with semi-permeable membranes in such a way that the volume accessible to each remains constant in the mixing process. In neither step is there any change of temperature, so that, using equation (8.13) for the entropy of the perfect gas, and using the fact that the entropy of the mixture is the sum of the entropies that the component gases would have if present alone, we have for the total entropy change

$$\Delta S = \Delta S_1 + \Delta S_2 = R\left[n_1 \ln\left(\frac{V_1+V_2}{V_1}\right) + n_2 \ln\left(\frac{V_1+V_2}{V_2}\right)\right] \quad (11.14)$$

where n_1 and n_2 are the numbers of moles of the gases present. As we should expect, ΔS is always positive.

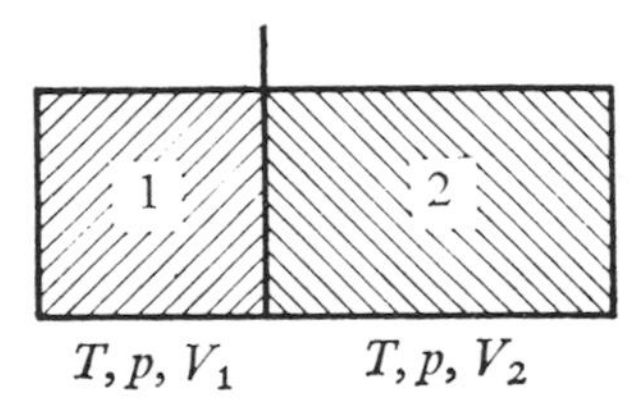

Fig. 11.2. Irreversible mixing of two gases.

Equation (11.14) leads to an interesting and apparently paradoxical result. It is no surprise that the diffusion of *different* gases leads to an increase in entropy, for this is clearly an irreversible process; but suppose that we initially fill the two volumes with the *same* gas. According to (11.14) there is still the same increase in entropy. But we know that, in this case, removing the partition sets in motion no irreversible process, as the composition is uniform throughout the vessel all the time, and therefore there cannot be any consequent change in entropy. This apparent contradiction is known as *Gibbs' paradox*. It is resolved when we realize that in constructing equation (11.14) we have assumed that the molecules of the two gases initially separated by the partition are *distinguishable*. For, in calculating the final entropy, we have used results we derived earlier for mixtures of *different* gases where we were able, by using semi-permeable membranes, to separate the mixture into its pure constituents. If the molecules are identical, such a separation cannot be made, and the arguments on which (11.14) is based break down. In principle, if we allow the gases to become more and more alike, we could always distinguish to which of the two a particular molecule belonged (although the process of doing so might become more and more difficult) and it would always be possible to make a device to separate them, *as long as they remained different*. But particles either are identical or they are not. There is no continuity from difference to identity; and there is no reason to expect a continuity of thermodynamic properties either. The sharp distinction between identical and non-identical (however similar) particles is, of course, fundamental in quantum theory and in statistical mechanics; but in the purely classical context in which it first arose, Gibbs' paradox presented a serious problem.

11.3. Chemical Potential

We shall now consider a general system of P phases and C components. We shall indicate phases by superscripts and components by subscripts.

The state of a single phase in thermodynamic equilibrium is determined by its temperature, pressure, and composition. Whereas the potentials of a pure substance are determined by three parameters, say the temperature, pressure, and total mass, we now have to specify the masses of each component separately. We may therefore write the Gibbs potential of the ϕth phase

$$G^{\phi} = G^{\phi}(T, p, m_1, m_2, \ldots m_C), \tag{11.15}$$

for which the differential form is

$$\begin{aligned} \mathrm{d}G^{\phi} &= \left(\frac{\partial G^{\phi}}{\partial T}\right)_{p,m_i} \mathrm{d}T + \left(\frac{\partial G^{\phi}}{\partial p}\right)_{T,m_i} \mathrm{d}p + \sum_{i=1}^{C} \left(\frac{\partial G^{\phi}}{\partial m_i}\right)_{T,p} \mathrm{d}m_i \\ &= -S^{\phi}\,\mathrm{d}T + V^{\phi}\,\mathrm{d}p + \sum_{i=1}^{C} g_i^{\phi}\,\mathrm{d}m_i. \end{aligned} \tag{11.16}$$

The quantities g_i^{ϕ} are known as the *chemical potentials* or *partial potentials* of the constituents in the ϕth phase. It is apparent from the form of (11.16) that they are intensive variables. Furthermore, they must be functions of state since they can only depend on the temperature, pressure, and composition of the phase.

Suppose, now, that we increase the total mass of a phase while keeping its temperature, pressure, and composition constant. Then we may write,

$$\mathrm{d}G^{\phi} = \sum_i g_i^{\phi}\,\mathrm{d}m_i^{\phi} = \sum_i g_i^{\phi}\, m_i^{\phi}\,\mathrm{d}\alpha \tag{11.17}$$

where $\mathrm{d}\alpha = \mathrm{d}m_i/m_i$ is the fraction by which the mass of each constituent is increased. Since the g_i^{ϕ} depend only on temperature, pressure, and composition, they must remain constant in this change so that (11.17) may be integrated to give

$$G^{\phi} = \sum_i g_i^{\phi} m_i^{\phi}. \tag{11.18}$$

Equation (11.18) shows that the chemical potentials are analogous to the specific Gibbs function of a pure substance. In a phase containing a single component, they are, of course, identical, but in mixtures, they depend always on concentration, and generally on the nature and proportions of the other components. Only in the limit that the components occupy the volume independently, as with mixtures of perfect gases, does the chemical potential of a component depend on temperature, total pressure, and the concentration of that component only. For this case, there are various useful forms of the partial potentials which follow directly from (11.11–13):

$$g_i(T, p_i) = g_{0i}(T) + \frac{RT}{M_{ri}} \ln p_i \tag{11.19}$$

$$g_i(T, p, c_i) = g_{0i}(T) + \frac{RT}{M_{ri}} \ln p + \frac{RT}{M_{ri}} \ln c_i \tag{11.20}$$

$$g_i(T, p, c_i) = g_{0i}(T, p) + \frac{RT}{M_{ri}} \ln c_i \tag{11.21}$$

where M_{ri} is the relative molecular mass. We have here dropped the phase superscript. For tidiness we shall usually drop the phase and component labels when the sense of an equation is obvious without them.

It should be noted that although we have defined the chemical potential through the Gibbs function, it is similarly related to the internal energy, enthalpy, and Helmholtz function. For example, by definition,

$$F(T, V, m_i) = G(T, p, m_i) - pV$$

and,

$$\begin{aligned} \mathrm{d}F &= \mathrm{d}G - p\,\mathrm{d}V - V\,\mathrm{d}p \\ &= -p\,\mathrm{d}V - S\,\mathrm{d}T + \sum g_i\,\mathrm{d}m_i. \end{aligned}$$

Therefore

$$\left(\frac{\partial F}{\partial m_i}\right)_{T,\,V} = g_i. \tag{11.22}$$

We therefore have the set of results:

$$\blacktriangleright \quad g_i=\left(\frac{\partial G}{\partial m_i}\right)_{T,\,p}=\left(\frac{\partial U}{\partial m_i}\right)_{S,\,V}=\left(\frac{\partial H}{\partial m_i}\right)_{S,\,p}=\left(\frac{\partial F}{\partial m_i}\right)_{T,\,V}. \qquad (11.23)$$

We may also derive Maxwell relations involving the chemical potentials in the same way as we did for simple one component systems. (Section 7.3.) Since G is a function of state, $\mathrm{d}G$ is an exact differential. Then, using the necessary and sufficient condition for G to be exact (section 1.9.3), we obtain[4]:

(*a*) *Relations between the chemical potentials and system variables*

$$\blacktriangleright \qquad \frac{\partial g_i}{\partial T}=-\frac{\partial S}{\partial m_i}=-s_i \qquad (11.24)$$

and

$$\blacktriangleright \qquad \frac{\partial g_i}{\partial p}=\frac{\partial V}{\partial m_i}=v_i \qquad (11.25)$$

where s_i and v_i are the *partial specific entropy* and the *partial specific volume* of the ith constituent: the quantities analogous to specific entropy and specific volume for a pure substance.

(*b*) *Symmetry relations between the chemical potentials*

$$\blacktriangleright \qquad \frac{\partial g_i}{\partial m_j}=\frac{\partial g_j}{\partial m_i}. \qquad (11.26)$$

We have used the m_i as the extensive component parameter to bring out the connection between the chemical potentials and the specific Gibbs function of a pure substance. Generally, it is more convenient to work in molar quantities and to use instead the number of moles of the components, n_i, rather than the masses. The differential form of G then becomes

$$\mathrm{d}G(T,p,n_i)=-S\,\mathrm{d}T+V\,\mathrm{d}p+\sum_{i=1}^{C}\mu_i\,\mathrm{d}n_i. \qquad (11.27)$$

The μ_i are known as the *molar partial potentials*. Clearly,

$$\mu_i=M_{ri}g_i \qquad (11.28)$$

where M_{ri} is the relative molecular mass of the ith component.

[4] The independent variables in G are T, p, m_1, m_2, . . . , m_C. For neatness, we do not list the constants in the partial differentials. It is implied that all the independent variables other than that appearing in the differential are held constant.

Results corresponding to those above follow in exactly the same way for the molar potentials:

$$G^{\phi}=\sum_{i}\mu_i^{\phi}n_i^{\phi} \tag{11.29}$$

▶ $$\mu_i=\left(\frac{\partial G}{\partial n_i}\right)_{T,\,p}=\left(\frac{\partial U}{\partial n_i}\right)_{S,\,V}=\left(\frac{\partial H}{\partial n_i}\right)_{S,\,p}=\left(\frac{\partial F}{\partial n_i}\right)_{T,\,V} \tag{11.30}$$

▶ $$\frac{\partial \mu_i}{\partial T}=-S_{mi}, \quad \text{the } \textit{partial molar entropy}, \tag{11.31}$$

▶ $$\frac{\partial \mu_i}{\partial p}=V_{mi}, \quad \text{the } \textit{partial molar volume}, \tag{11.32}$$

▶ $$\frac{\partial \mu_i}{\partial n_j}=\frac{\partial \mu_j}{\partial n_i} \tag{11.33}$$

and for perfect gas mixtures,

$$\mu_i(T,p,n_i)=\mu_{0i}(T)+RT\ln p+RT\ln c_i \tag{11.34}$$

$$\mu_i(T,p,n_i)=\mu_{0i}(T,p)+RT\ln c_i. \tag{11.35}$$

11.4. Conditions for Equilibrium

The condition for thermodynamic equilibrium of any system at constant temperature and pressure is that its Gibbs function be a minimum. Generally, it is not practicable to try to write down a general expression for the total Gibbs potential and then to minimize it to find the equilibrium configuration; but instead one applies $dG=0$ to all possible infinitesimal displacements of the system away from equilibrium. For each possible displacement, this places a condition on the system variables from which useful results follow. This is the approach we adopted in chapter 10 in discussing change of phase in one component systems. The only difference which the introduction of several components makes is that the Gibbs function now contains a greater number of variables, and a correspondingly greater number of displacements is possible. Exactly what displacements are possible in any particular case depends on the detailed nature of the system and on any constraints which may be effective. For example, a semipermeable membrane between two phases may permit the transfer from one phase to the other of one component while preventing that of the others. Similarly, if no chemical reactions may take place, the total mass of each component must be conserved, whereas if there is chemical reaction, the masses of the reacting components may be varied in accordance with the reaction involved.

Let us first of all assume that we have a multicomponent, multiphase system in which all phases are in direct contact, and in which there are no surface effects or chemical reactions. Then the only displacements accessible to the system are those involving the transfer of mass from one phase to another. Suppose, for example, that we transfer $\mathrm{d}n_i$ moles of the ith constituent from phase 1 to phase 2. Since G is a minimum against any displacement, it must be to this displacement in particular. Therefore

$$\mathrm{d}G = -\mu_i^1\,\mathrm{d}n_i + \mu_i^2\,\mathrm{d}n_i = 0$$

or

$$\mu_i^1 = \mu_i^2.$$

The argument may be applied to any two phases to give the set of equations

$$\mu_i^1 = \mu_i^2 = \ldots = \mu_i^P. \tag{11.36}$$

The argument may also be repeated for each component so that there are C sets of equations like (11.36):

▶ $$\mu_i^1 = \mu_i^2 = \ldots = \mu_i^P \quad \textit{for all } i. \tag{11.37}$$

In terms of the chemical potentials, the corresponding result is

$$g_i^1 = g_i^2 = \ldots = g_i^P \quad \textit{for all } i. \tag{11.38}$$

The condition for equilibrium between phases of a single component system which we derived in chapter 10 is seen to be contained in these results as the special case when there is only one component present and the chemical potential becomes identical to the specific Gibbs function.

Now consider a possible chemical reaction in phase ϕ. Suppose that this has the form

$$a_1J_1 + a_2J_2 \rightleftharpoons a_3J_3 + a_4J_4 \tag{11.39}$$

where the J's are the reacting constituents and the a's are the numbers of molecules of each taking part in the reaction. For convenience, we will rewrite this with all reactants on the same side of the equality:

$$-a_1J_1 - a_2J_2 + a_3J_3 + a_4J_4 = 0$$

or, changing the symbols to achieve a uniform notation,

$$\nu_1J_1 + \nu_2J_2 + \nu_3J_3 + \nu_4J_4 = 0 \tag{11.40}$$

where it is implied that the ν's for the species consumed when the reaction proceeds in the forward direction are negative ($\nu_1 = -a_1$, etc.).

Now, for equilibrium we must have $\mathrm{d}G = 0$ for an infinitesimal displacement of the reaction. Suppose that we displace it in the forward direction so that $-\nu_1\,\mathrm{d}\alpha$ moles of J_1 and $-\nu_2\,\mathrm{d}\alpha$ moles of J_2 are consumed and $+\nu_3\,\mathrm{d}\alpha$ moles of J_3 and $+\nu_4\,\mathrm{d}\alpha$ moles of J_4 produced. That is, the numbers of moles of all reactants increase by $\nu_i\,\mathrm{d}\alpha$. Then

$$\mathrm{d}G = \mathrm{d}G^{\phi} = (\mu_1^{\phi}\nu_1 + \mu_2^{\phi}\nu_2 + \mu_3^{\phi}\nu_3 + \mu_4^{\phi}\nu_4)\,\mathrm{d}\alpha = 0, \qquad (11.41)$$

which gives, dropping the phase superscript,

$$\mu_1\nu_1 + \mu_2\nu_2 + \mu_3\nu_3 + \mu_4\nu_4 = 0 \qquad (11.42)$$

which is known as the *equation of reaction equilibrium.* It clearly generalizes for reactions involving any number of reactants to

$$\blacktriangleright \qquad \sum \mu_i\,\nu_i = 0. \qquad (11.43)$$

From these results, we may derive a simple expression for the number of degrees of freedom which a system has in terms of the number of components present, the number of chemical reactions which may occur and the number of phases we require to coexist. (We shall not consider other possible restrictions.) *Ab initio*, the system has $2 + CP$ degrees of freedom:

$$\begin{array}{llll} T, p, m_1^1, & m_2^1, & \ldots, & m_C^1 \\ \quad\quad m_1^2, & m_2^2, & \ldots, & m_C^2 \\ \quad\quad \cdot & \cdot & & \cdot \\ \quad\quad \cdot & \cdot & & \cdot \\ \quad\quad \cdot & \cdot & & \cdot \\ \quad\quad m_1^P, & m_2^P, & \ldots, & m_C^P. \end{array}$$

The C sets of equations (11.37) contain $C(P-1)$ conditions, and each chemical reaction introduces a further condition of the form (11.43). Suppose that there are R chemical reactions. Then the number of degrees of freedom which remain to the system are

$$\mathcal{N}_{\text{total}} = 2 + CP - C(P-1) - R = 2 + C - R. \qquad (11.44)$$

However, we are not usually concerned with the *amount* of each phase present but only with their composition and number. We are normally only concerned with those degrees of freedom which are *intensive.* Clearly, the extent of each phase must correspond to one degree of freedom so that from the total number we have calculated we must remove a further P to obtain the number of intensive degrees of freedom :

$$\blacktriangleright \qquad \mathcal{N}_{\text{Int.}} = 2 + C - P - R. \tag{11.45}$$

This result is known as the *Phase Rule*[5]. We see that this is in accord with our earlier discussion of single component systems. With one phase, we have two degrees of freedom (e.g. p and T); if two phases coexist, there is one (whence the Clausius–Clapeyron equation); if three phases are to coexist, there is no freedom (the triple point); and if a substance exists in more than three phases under no condition may more than three phases coexist at equilibrium (e.g., He^4, Fig. 10.13).

11.5. Ideal Solutions

In order to discuss the behaviour of solutions, we need to express the partial potentials of the components in a form in which the concentrations appear explicitly. We first derive expressions which hold for dilute solutions.

It is physically obvious, and also an experimental fact, that for a sufficiently dilute solution, the partial pressure of the solute in the *vapour* phase must be proportional to its concentration in the solution. This is known as *Henry's law*. Then the partial potential of the solute *in the vapour phase* must obey

$$\left(\frac{\partial \mu_i}{\partial c_i}\right)_{T,\,p} = \left(\frac{\partial \mu_i}{\partial p_i}\right)_{T,\,p} \left(\frac{\partial p_i}{\partial c_i}\right)_{T,\,p} = \frac{RT}{c_i} \tag{11.46}$$

where we have used the perfect gas law for the solute vapour, and c_i is the concentration of the solute *in the solution.* Integrating,

$$\mu_i = \mu_{0i}(T, p) + RT \ln c_i. \tag{11.47}$$

But, for equilibrium, the partial potentials for each component in the liquid and vapour phases must be equal, so that that of the solute

[5] Here we have taken the number of components as the total number of distinguishable chemical constituents. Sometimes, the number of components is taken as the smallest number of constituents which is necessary to determine the composition of all phases. This is essentially C–R. The number of independent reactions does not then appear explicitly in the phase rule which takes the form $\mathcal{N}_{\text{Int.}} = 2 + C' - P$.

in the solution is also given by (11.47). This argument applies to all solute species, so that for dilute solutions we have

$$\mu_i = \mu_{0i}(T, p) + RT \ln c_i \qquad \textit{for all solutes.} \qquad (11.48)$$

We may also show that for dilute solutions the molar partial potential of the *solvent* takes the same form. We start with the pure solvent and add each solute in turn. Consider first adding the jth solute. Then with (11.6), (11.48) becomes

$$\mu_j = \mu_{0j} + RT \ln n_j - RT \ln (n_j + n_s) \qquad (11.49)$$

where the subscript s indicates the solvent. To obtain μ_s we use the symmetry relation (11.33) in the form :

$$\frac{\partial \mu_s}{\partial n_j} = \frac{\partial \mu_j}{\partial n_s}.$$

Then, from (11.49),

$$\frac{\partial \mu_j}{\partial n_s} = -\frac{RT}{n_j + n_s} = \frac{\partial \mu_s}{\partial n_j}.$$

Integrating with respect to n_j,

$$\mu_s(n_j) - \mu_s(0) = -RT \ln \left(\frac{n_j + n_s}{n_s}\right)$$

$$\simeq RT \ln (1 - c_j). \qquad (11.50)$$

Repeating the integration for successive solvents, we obtain

$$\mu_s(n_j, n_k, \ldots) - \mu_s(0, 0, \ldots) = RT \sum_i \ln (1 - c_i)$$

$$= RT \ln \left[\prod_i (1 - c_i)\right]$$

$$\simeq RT \ln \left(1 - \sum_i c_i\right);$$

or, changing the notation,

$$\mu_s = \mu_{0s} + RT \ln \left(1 - \sum_k c_k\right) = \mu_{0s} + RT \ln c_s \qquad (11.51)$$

where μ_{0s} is the molar partial potential of the pure solvent. Thus for dilute solutions,

► $$\mu_i = \mu_{0i}(T, p) + RT \ln c_i \qquad \textit{for all components} \qquad (11.52)$$

although normally it is only for the solvent that μ_0 is the molar partial potential of the pure species. With this one difference, the equations of (11.52) are seen to be identical to those for the molar partial potentials of perfect gas mixtures (equations (11.35)). This is not surprising, for the physical approximations involved are similar: Henry's law is only true when the solute atoms are far enough apart for their mutual interactions to be negligible. When this is the case we would expect them to behave like a perfect gas, but one whose properties are modified by the presence of the solvent.

For certain solutions (11.52) hold well for all concentrations. These are known as *ideal solutions*. In this case, it is apparent by taking the limit as $c_i \to 1$, that the μ_{0i} are the molar partial potentials of the pure species *for all components*. Generally, solutions whose behaviour approaches the ideal contain chemically similar components. For example, ethylene bromide and propylene bromide are miscible in all proportions to give solutions which are very nearly ideal.

In non-ideal solutions, the μ_i may vary in such a way that it becomes energetically favourable for mixtures of certain compositions to separate into two phases of different composition. This will occur when the total Gibbs potential for the two phase configuration is less than that of the single phase (the homogeneous mixture). Such is the case with mixtures of methyl alcohol and carbon disulphide. At room temperature, the maximum solubility of methanol in carbon disulphide is 2·5 per cent by weight and of carbon disulphide in methanol, 50 per cent by weight. Mixtures with between 2·5 and 50 per cent of methanol separate into two phases, one containing 2·5 per cent and the other 50 per cent. The phase diagram for mixtures of methanol and carbon disulphide therefore has a region of *heterogeneous equilibrium* in which more than one phase is present. The Gibbs potential is lower in the heterogeneous configuration than in the homogeneous. Similar considerations determine the phase diagrams in more complicated situations where solid and gaseous phases may also be present; but a discussion of these would lead us too far afield[6].

11.5.1. The Vapour Pressure of an Ideal Solution

Consider an ideal solution in contact with its vapour which we shall assume to behave as a mixture of perfect gases. Then using (11.52), (11.35), and (11.7), we may put the condition that there is

[6] For a more detailed discussion, see A. H. Wilson, *Thermodynamics and Statistical Mechanics*, chapter 12, Cambridge University Press, 1957; and J. C. Slater, *Introduction to Chemical Physics*, chapter XVII, McGraw-Hill, 1939.

equilibrium between the liquid and vapour phases in the form

$$\mu_{0i}^{L}(T, p) + RT \ln c_i^L = \mu_{0i}^{G}(T, p_i^0) + RT \ln \frac{p_i}{p_i^0} \tag{11.53}$$

for all components, where p, p_i^0, and p_i are the total pressure, the vapour pressure of the pure constituent and its partial pressure in the vapour, and the superscripts L and G refer to the liquid and gaseous phases. (In effect, we have used p_i^0 as a reference pressure for the vapour.) If we let $c_i \to 1$, then $p \to p_i^0$ and (11.53) becomes

$$\mu_{0i}^{L}(T, p_i^0) = \mu_{0i}^{G}(T, p_i^0). \tag{11.54}$$

Eliminating μ_{0i}^{G} between (11.53) and (11.54) and rearranging,

$$\mu_{0i}^{L}(T, p) - \mu_{0i}^{L}(T, p_i^0) = RT \ln \frac{p_i}{c_i^L p_i^0}. \tag{11.55}$$

Now the density of the liquid is normally much greater than that of the vapour, so that whether p is very different from p_i^0 or not, the left-hand side of (11.55) is certainly much smaller than RT. We may therefore equate the logarithm to zero, which gives

► $$p_i = c_i^L p_i^0 \qquad \textit{for all components.} \tag{11.56}$$

This set of equations is known as *Raoult's law*. This differs from Henry's law in that it applies to *all components at all concentrations*. The proportionality stated in Henry's law applies only for *solutes* at *low* concentration, and, of course, it is only for ideal solutions that the constant of proportionality is equal to the natural vapour pressure of the solute.

From (11.56) it follows that the total vapour pressure of a perfect solution is simply an average of the natural vapour pressures of the components weighted according to their concentrations *in the solution*:

$$p = \sum c_i p_i^0. \tag{11.57}$$

11.5.2. Solubility in Ideal Solutions

It is sufficient to demonstate the physical principles if we consider a simple pure solute i in contact with its solution. Then the condition for the solution to be saturated is

$$\mu_i^S(T, p) = \mu_i^L(T, p) \tag{11.58}$$

where S and L indicate the pure solute and solution (L = liquid) respectively. For ideal solutions, μ_i^L is independent of the nature of the other components, so that the solubility of a particular solute will be the same function of temperature and pressure for all solvents in which it forms ideal solutions. Now, if we change the temperatures, keeping the solution saturated,

$$\mathrm{d}\mu_i^{\mathrm{S}} = \mathrm{d}\mu_i^{\mathrm{L}}. \tag{11.59}$$

If the pressure is constant in the change,

$$\mathrm{d}\mu_i^{\mathrm{L}} = \left(\frac{\partial \mu_i^{\mathrm{L}}}{\partial T}\right)_{p,\,c_i} \mathrm{d}T + \left(\frac{\partial \mu_i^{\mathrm{L}}}{\partial c_i}\right)_{p,\,T} \mathrm{d}c_i. \tag{11.60}$$

Substituting from (11.31) and (11.52),

$$\mathrm{d}\mu_i^{\mathrm{L}} = -S_{mi}^{\mathrm{L}}\,\mathrm{d}T + RT\,\frac{\mathrm{d}c_i}{c_i} \tag{11.61}$$

where S_{mi}^{L} is the molar entropy of the solute in the liquid. Also, at constant pressure,

$$\mathrm{d}\mu_i^{\mathrm{S}} = -S_{mi}^{\mathrm{S}}\,\mathrm{d}T \tag{11.62}$$

where S_{mi}^{S} is the partial molar entropy of the pure solute.
From (11.61) and (11.62)

$$\frac{\mathrm{d}c_i}{c_i} = (S_{mi}^{\mathrm{L}} - S_{mi}^{\mathrm{S}})\,\frac{\mathrm{d}T}{RT} = L\,\frac{\mathrm{d}T}{RT^2} \tag{11.63}$$

where $L = T(S_{mi}^{\mathrm{L}} - S_{mi}^{\mathrm{S}})$ is the *molar heat of solution*. (The heat *absorbed* when one mole passes into solution.) If the solute is a solid, then for moderate temperature intervals the variation in L is relatively small, and (11.63) may be integrated to give an approximate explicit form for c_i:

▶ $$c_i = c_i^0 \exp(-L/RT) \tag{11.64}$$

11.5.3. Osmotic Pressures

For ideal solutions, the addition of a solute to a solvent reduces the chemical potential of the solvent. If, then, a pure solvent is separated from a solution by a wall permeable to the solvent, equilibrium will not exist, but the solvent will pass through into the solution. This flow may be stopped by applying an excess hydrostatic pressure to the solution so as to raise the chemical potential of the solvent in it to equal that of the pure species. The excess pressure which must be applied is known as the *osmotic pressure* and the equilibrium so obtained as *osmotic equilibrium*.

In osmotic equilibrium, the osmotic pressure Π must satisfy

$$\mu_s(T, p+\Pi, c_i)=\mu_s(T, p, 0) \tag{11.65}$$

where μ_s is the molar partial potential of the solvent and the c_i are the concentrations of the solutes. For ideal solutions this becomes

$$RT \ln (1-\Sigma c_i)=\mu_{0s}(T, p)-\mu_{0s}(T, p+\Pi)$$
$$=-\int_p^{p+\Pi} V_{ms}(T, p')\, dp'$$

where μ_{0s} and V_{ms} are the molar partial potential and the molar volume of the pure solvent. If the liquid is incompressible, this becomes

$$\Pi V_{ms}=-RT \ln (1-\Sigma c_i)=-RT \ln c_s \tag{11.66}$$

which, if the solution is dilute, reduces further to

▶ $$\Pi V_{ms}=RT\, \Sigma c_i \tag{11.67}$$

where the summation is over the solutes.

Formally, this equation is similar to that for the total pressure of a mixture of perfect gases. Osmotic pressures, however, may be very large since solutions of moderate concentration will correspond to gases at densities approaching that of the liquid state.

11.6. Ideal Gas Reactions

Consider a simple ideal gas reaction of the form

$$\Sigma \nu_i J_i=0. \tag{11.68}$$

For this, the equation of reaction equilibrium, (11.43), is

$$\Sigma \nu_i \mu_i=0. \tag{11.69}$$

For a mixture of ideal gases, the μ_i are, according to equation (11.34),

$$\mu_i=\mu_{0i}(T)+RT \ln p_i \tag{11.70}$$

where the μ_{0i} are the molar partial potentials at temperature T and at the reference pressure $p_i=1$. Substituting in the equation of reaction equilibrium we obtain

$$\Sigma \nu_i \mu_{0i}+RT \ln \Pi p_i^{\nu_i}=0$$

or

▶ $$\Pi p_i^{\nu_i}=K_p(T) \tag{11.71}$$

where

▶ $$\ln K_p(T)=-\frac{1}{RT} \Sigma \nu_i \mu_{0i}. \tag{11.72}$$

Equation (11.71) is known as the *law of mass action* and K_p as the *equilibrium constant.* (The suffix is included to indicate that it is defined through the partial pressures of the reacting gases rather than through their concentrations or partial volumes which give alternative forms.)

These results are of fundamental importance in physical chemistry. The law of mass action itself determines how the equilibrium concentrations of reacting gases are related for a given temperature, while through the behaviour of the equilibrium constant it is possible to derive thermodynamic information about the reactants. For example, K_p is simply related to the *heat of reaction* Q^*, defined as the heat *absorbed* when the reaction proceeds in the forward direction by ν_i moles of each constituent. (For example, in the simple reaction $a_1J_1+a_2J_2 \rightleftharpoons a_3J_3+a_4J_4$, Q^* is the heat absorbed when a_1 moles of J_1 react with a_2 moles of J_2 to produce a_3 moles of J_3 and a_4 moles of J_4.) We derive the relation between K_p and Q^* as follows:

From (11.72) and using (11.31), (11.19), and (11.5),

$$RT^2 \frac{\mathrm{d}}{\mathrm{d}T} \ln K_p = \Sigma\, \nu_i(\mu_{0i}+T\,S_{m0i})$$

$$= \Sigma\, \nu_i\{(\mu_{0i}+RT \ln p_i)+T\,(S_{m0i}-R \ln p_i)\}$$

$$= \Sigma\, \nu_i\,(\mu_i+TS_{mi})$$

$$= \Sigma\, \nu_i\, H_{mi}$$

where the H_{mi} are the molar enthalpies of the reactants.[7] We may rewrite this more simply as

$$RT^2 \frac{\mathrm{d}}{\mathrm{d}T} \ln K_p = \Delta H_m. \tag{11.73}$$

This is a famous equation of chemical thermodynamics known as the *van't Hoff isobar.*

Now, if we imagine that the reaction occurs in a box equipped with a set of semipermeable membranes through which the reactants are supplied and the products emerge, all at their appropriate partial pressures (Fig. 11.3), then the reaction proceeds as a simple flow process and the energy entering as heat from the surroundings is

[7] Alternatively, we may remember that the enthalpy of a perfect gas is independent of pressure and put $\mu_{0i}+TS_{m0i}=H_{m0i}=H_{mi}$.

the difference of the enthalpies of the gases leaving and entering the box. Thus

$$Q^* = \Sigma\, \nu_i\, H_{mi} = \Delta H_m$$

and we have

$$RT^2 \frac{\mathrm{d}}{\mathrm{d}T} \ln K_p = \Delta H_m = Q^* \tag{11.74}$$

which is the relation between the equilibrium constant and the heat of reaction.

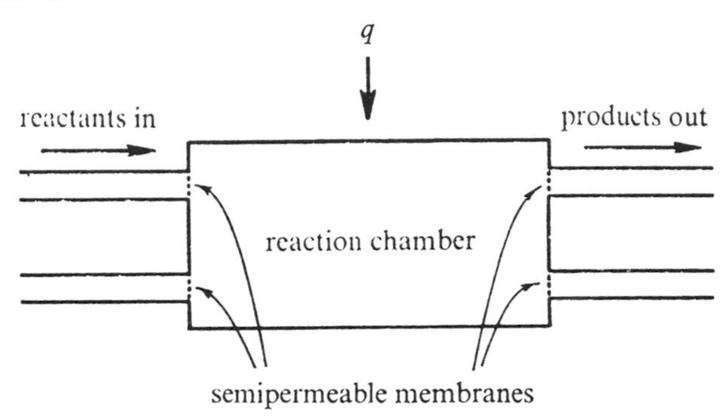

Fig. 11 . 3. A gas reaction as a simple flow process.

The law of mass action has a straightforward microscopic interpretation. The simplest microscopic assumption that we can make is that reaction can only occur when the molecules required for it are close together. For example, in the reaction

$$a_1 J_1 + a_2 J_2 \rightleftharpoons a_3 J_3 + a_4 J_4$$

for a_1 molecules of J_1 to combine with a_2 molecules of J_2 to give the products on the right we may suppose that we require a 'collision' involving a_1 molecules of J_1 and a_2 molecules of J_2. Now, the probability that *one* molecule of J_1 will be in a particular region at a particular time will be proportional to the density of J_1 molecules which is, in turn, proportional to the partial pressure of J_1. Thus, the probability that a_1 molecules of J_1 coincide is proportional to $p_1^{a_1}$, and the probability of a collision event of the kind necessary for reaction from left to right is proportional to $p_1^{a_1} p_2^{a_2}$. The *rate* of reaction from left to right will therefore be

$$C(T)\, p_1^{a_1}\, p_2^{a_2}$$

where the constant of proportionality takes account of the microscopic properties of the reactants and of the kinetics of the collision.

Similarly, the rate of reaction to the left will be

$$C'(T)\, p_3^{a_3}\, p_4^{a_4}$$

and at equilibrium these will be equal, giving

$$\frac{p_1^{a_1}\, p_2^{a_2}}{p_3^{a_3}\, p_4^{a_4}} = K(T)$$

which is the law of mass action.

12. The Third Law

12.1. The Third Law

The third law is concerned with the limiting behaviour of systems as the temperature approaches absolute zero. The statement of the law in the form due to *Simon* reads:

The contribution to the entropy of a system by each aspect which is in internal thermodynamic equilibrium tends to zero as the temperature tends to zero. By *aspect* of a system is meant a part of the system or a process in it which interacts only weakly with the rest of the system and therefore makes an essentially independent contribution to the properties of the whole. Thus, in our discussion of paramagnetic salts (sections 5.6.2 and 8.8.1) we were able to speak of separate contributions to the entropy from the thermal motions of the lattice and from the disorder of the magnetic subsystem. According to the third law, both these contributions will go to zero as the temperature goes to zero provided that both subsystems are in thermodynamic equilibrium. It should be noted that the third law does not exclude states for which the equilibrium is metastable. As we shall see later, its application to metastable systems provides some of the best experimental evidence for its truth.

Experiments can only determine *differences* in entropy, and in its earlier forms the third law stated only that the entropy due to each aspect of all systems took the *same* value at absolute zero. The choice of zero for this universal constant brings the third law into agreement with the Boltzmann relation (equation (5.19)):

$$S = k \ln g$$

for in quantum mechanics there is an unproved theorem which states that the ground state of any system is non-degenerate, so that in the ground state there is complete order, $g=1$ and $S=0$. The status of this theorem of quantum mechanics is similar to that of the third law in thermodynamics. However, we must remember that in deriving the Boltzmann relation itself (section 8.3), we *chose* to make the constant of integration zero. The essential point of the third law is that the constant is the same for all systems, and it must be stressed that it is strictly only a matter of convenience to set it equal to zero.

Various attempts have been made to derive the third law from the second; but, in fact, this cannot be done. Its status is that of an independent fundamental postulate[1].

In the following sections we shall discuss some of the consequences and applications of the third law, restricting ourselves for the greater part to the use of classical ideas but invoking also simple microscopic and quantal concepts when relevant. A full appreciation of the significance of the third law requires also a discussion in the context of quantum statistical mechanics which should be sought elsewhere[2].

12.2. Elementary Physical Consequences of the Third Law

A most important consequence of the third law is that all heat capacities must tend to zero as the temperature approaches zero. This may be seen by writing the heat capacity in the form

$$C_x = T\left(\frac{\partial S}{\partial T}\right)_x = \left(\frac{\partial S}{\partial \ln T}\right)_x. \tag{12.1}$$

As $T \to 0$, $\ln T \to -\infty$ and $S \to 0$, so that the derivative also tends to zero. We may reach the same conclusion by considering the total change of entropy on cooling from T to absolute zero:

$$\Delta S = \int_T^0 \frac{C}{T}\,\mathrm{d}T.$$

The integral is only finite if $C \to 0$ as $T \to 0$. Thus, *all heat capacities tend to zero as the temperature approaches absolute zero.* This result emphasizes the connection between the third law and quantum theory, for classical heat capacities do not vary with temperature. (Equipartition of energy.) It is therefore impossible to construct a classical interpretation of the third law. In fact, as we have previously pointed out (section 8.2.4) quantum theory demands that eventually all specific heats vanish exponentially.

Through the Maxwell relations, the derivatives of the entropy are related to derivatives of other system parameters. With the third

[1] See, A. B. Pippard, *Classical Thermodynamics*, Cambridge University Press, pp. 48–51, 1966.

[2] For example, J. Wilks, *The Third Law of Thermodynamics*, Oxford University Press, 1961, and A. H. Wilson, *Thermodynamics and Statistical Mechanics*, Cambridge University Press, 1957.

law these lead to conditions on the limiting behaviour of other common thermodynamic quantities.

For example, for a system subject to work by hydrostatic pressure, we have

$$\left(\frac{\partial S}{\partial p}\right)_T = -\left(\frac{\partial V}{\partial T}\right)_p = -V\beta_p$$

where β_p is the isobaric cubic expansivity. By the third law

$$\underset{T\to 0}{\mathrm{Lt}}\left(\frac{\partial S}{\partial p}\right)_T = 0$$

so that

$$\beta_p \to 0. \tag{12.2}$$

Corresponding results follow for other properties. Where surface energies are important,

$$\left(\frac{\partial S}{\partial A}\right)_T = -\frac{\mathrm{d}\sigma}{\mathrm{d}T}$$

so that by the third law, surface tension becomes constant:

$$\frac{\mathrm{d}\sigma}{\mathrm{d}T} \to 0. \tag{12.3}$$

This is found to be so for both liquid ^{4}He and liquid ^{3}He, the only liquids which exist at absolute zero.

Again, for a paramagnetic substance,

$$\left(\frac{\partial S}{\partial B}\right)_T = \left(\frac{\partial m}{\partial T}\right)_B = \frac{VB}{\mu_0}\left(\frac{\partial X_m}{\partial T}\right)_B,$$

so that by the third law

$$\left(\frac{\partial \chi_m}{\partial T}\right)_B \to 0. \tag{12.4}$$

This is an interesting result because we see that the third law requires Curie's law ($\chi_m = a/T$) to break down at some temperature. Again, we know from quantum theory that this must be the case, for Curie's law corresponds to the classical limit in which the spacings of the energy levels of the magnetic subsystems are small compared with thermal energies, and since the ground state must be separated from the next nearest state by a finite energy difference (as is required by its non-degeneracy) the classical limit cannot hold to absolute zero.

The Clausius–Clapeyron equation (10.18) gives the gradient of the boundary between two phases of a pure substance:

$$\frac{\mathrm{d}p}{\mathrm{d}T}=\frac{\Delta S}{\Delta V}.$$

According to the third law, $\Delta S \to 0$ as $T \to 0$, so that in this limit the phase boundary must become parallel to the T-axis. The only substance on which careful experiments have been made is ^{4}He, for which solid and liquid may coexist at absolute zero. These have shown that the gradient of the melting curve does approach zero at the lowest temperatures. (See Fig. 10.13.) At first sight, it is surprising that the entropy of a liquid can ever vanish, for one normally thinks of a liquid as being necessarily less ordered than a solid. The explanation is that the entropy of the system does not only depend on the positions of the atoms but also on their momenta, and for precisely the same reason as helium may remain liquid to absolute zero (section 10.9.2), the state of zero entropy of the liquid corresponds to a configuration in which the ordering is dominant in *momentum* rather than *position*. Again, this is a consequence of quantum effects. A similar argument must apply to ^{3}He which can also exist as a liquid at absolute zero. The zero entropy configuration must again be one in which there is ordering in the momenta of the particles; but, of course, there is no reason why the ground state of ^{3}He should show extraordinary properties like superfluidity. However, ^{3}He is interesting for other reasons. Below about 0·3 K the slope of the melting curve becomes negative (Fig. 12.1), the solid becoming the high temperature phase. According to the Clausius–Clapeyron equation this means that either ΔV or ΔS becomes negative. Since the change is not associated with any singularity, it is unlikely to be ΔV and, in fact, measurements show that this is nearly constant. It therefore follows that below 0·3 K the entropy of the solid is greater than that of the liquid! The reason for this is that the helium nucleus, consisting of two protons and one neutron, has a net spin (and associated magnetic moment). Now the entropy must, of course, include a contribution related to the degree of order in the spin system. Magnetic measurements show that as the liquid is compressed towards solidification, it obeys Curie's law to lower temperatures, implying that magnetic ordering sets in at higher temperatures at liquid densities than it does at solid. This is presumably because the uncertainty in atomic position in the liquid allows a stronger interaction between the nuclear moments.

Thus for a certain range of temperatures the liquid is more ordered than the solid and the gradient of the melting curve becomes negative. Eventually, of course, it must again become zero in accordance with the third law.

A relation analogous to the Clausius–Clapeyron equation describes the variation of the critical field of a superconductor:

$$\frac{dB_c}{dT} = -\frac{\Delta s}{\Delta m} \qquad \text{(equation (10.39))}$$

so that the third law requires the critical field to become constant as the temperature approaches absolute zero.

All these elementary consequences of the third law are well supported by experiments.

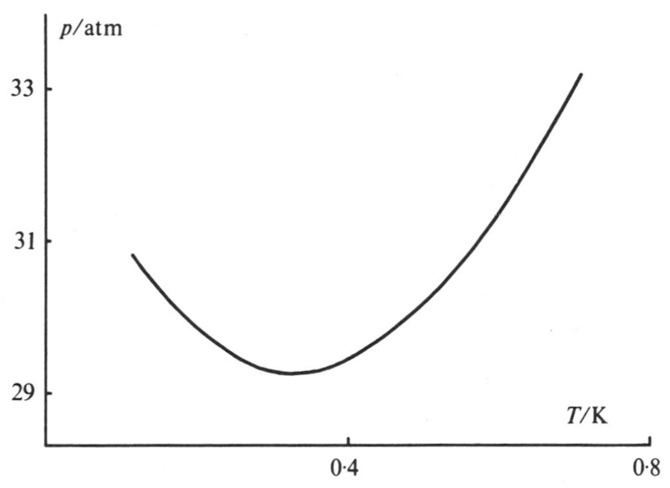

Fig. 12 . 1. Part of the melting curve of ^{3}He.[3]

12.3. The Unattainability of Absolute Zero

It follows from the third law that it is impossible to cool a system to absolute zero. The result may be stated formally as follows:

It is impossible to reduce the temperature of any system or part of a system to the absolute zero in a finite number of operations.

This is known as the *unattainability statement of the third law* and is sometimes used as an alternative to the Simon statement, but it is not exactly equivalent to it as we shall now show.

Consider a system in which we are attempting to produce cooling

[3] J. L. Baum, D. F. Brewer, J. G. Daunt, and D. O. Edwards, *Phys. Rev. Lett.*, **3**, 127, 1959.

by varying a parameter X. (If we are concerned with cooling by adiabatic demagnetization, X might be the applied field.) Suppose that by varying X from X_1 to X_2 we cause the system to cool from T_1 to T_2. Then, using the second law but not the third, we may write the entropies of the initial and final states

$$S(T_1, X_1) = S(0, X_1) + \int_0^{T_1} \left(\frac{\partial S}{\partial T}\right)_{X=X_1} \mathrm{d}T$$

and

$$S(T_2, X_2) = S(0, X_2) + \int_0^{T_2} \left(\frac{\partial S}{\partial T}\right)_{X=X_2} \mathrm{d}T.$$

Since heat capacities are always positive[4], we will achieve the greater cooling if we make the final entropy as small as possible. This requires the process to be adiabatic (heat could only leak *in* if we are trying to cool below the temperature of the surroundings) and reversible. In that case,

$$S(0, X_1) + \int_0^{T_1} \left(\frac{\partial S}{\partial T}\right)_{X=X_1} \mathrm{d}T = S(0, X_2) + \int_0^{T_2} \left(\frac{\partial S}{\partial T}\right)_{X=X_2} \mathrm{d}T. \quad (12.5)$$

Assuming the truth of the Simon statement of the third law,

$$S(0, X_1) = S(0, X_2)\ (=0),$$

and for T_2 to be zero,

$$\int_0^{T_1} \left(\frac{\partial S}{\partial T}\right)_{X=X_1} \mathrm{d}T = 0.$$

But, except at absolute zero, $(\partial S/\partial T)_X > 0$[4] so that there is no non-zero solution for T_1. Thus the unattainability of absolute zero follows from the Simon statement of the third law. The connection is illustrated in Fig. 12.2.

[4] This is obvious from the quantum point of view since, however great the energy separation between the ground state and the next lowest state of the system, increasing the temperature necessarily raises the average energy even though by a very small amount. From the thermodynamic point of view it is sufficient to note that $(\partial S/\partial T)_X < 0$ is impossible for reasons of stability. $(\partial S/\partial T)_X = 0$ is also inadmissible since if the system can exchange no heat the whole concept of temperature breaks down. Thus, for any real system $(\partial S/\partial T)_X > 0$ at $T > 0$.

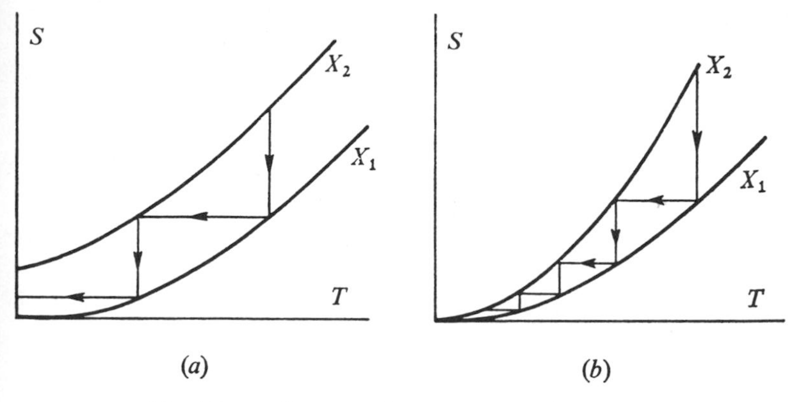

Fig. 12 . 2. Hypothetical entropy diagrams.

The system represented in (*a*) violates the third law and may be cooled to absolute zero in a finite number of operations. The third law is satisfied by (*b*) and absolute zero cannot be reached in a finite number of steps.

If we try to reverse the argument to prove the Simon statement by applying the unattainability statement to (12.5) we only succeed in establishing that

$$S(0, X_1) = S(0, X_2).$$

This is clearly in agreement with the Simon statement but is less than it, since it does not prove that the entropies of *all* systems or parts of systems are equal at absolute zero. Nor, of course, does it place the entropy equal to zero. As we have already stressed, this latter part of the Simon statement is only a matter of convenience and is not an important part of its substance; but the universal aspect of the Simon statement is important. Its verification lies in the success with which it is applied to chemical systems. (See section 12.6.)

12.4. Allotropic Transformations

Many solids may exist in more than one stable crystalline form. This often happens when the energies of the different structures lie close together so that as the external conditions are varied, one or another may become the most stable. Such changes of structure are known as *allotropic transformations*, and we have already referred to those in iron as examples of first order change of phase (section 10.9.1). In many cases, on cooling through the transition temperature, the transformation takes place rapidly and very little supercooling of the high temperature phase can be achieved. When this is the case the thermal energies must be comparable to any potential barrier separating the two configurations so that the rearrangement proceeds rapidly. In some cases, however, the transformation is slow, and by cooling rapidly through the transformation temperature, the high temperature phase may be retained indefinitely as a well defined metastable state. At absolute zero, the third law will apply both to the low temperature phase and also to the metastable allotrope. (It matters only that the configuration corresponds to *a* minimum in the energy, for at absolute zero this results in a well defined stable state since there can be no thermal excitation over the potential barrier which must separate it from any lower minimum[5].) The entropy of the high temperature phase at the transition temperature may then be calculated by two paths: (1) by integration of C/T for the metastable allotrope (which is the high temperature phase) from absolute zero to the transition temperature; and (2) by integration of C/T for the low temperature phase from absolute zero to the transition temperature and adding the entropy change in the transition, L/T. If the third law is true then the same value should be found in both calculations.

The third law has been tested in this way by measurements on several solids showing suitable allotropic transformations. These include tin[6], cyclohexanol[7], and sulphur[8], but perhaps the most interesting are the measurements on phosphine[9] which has four allotropic forms. The transitions in phosphine are indicated in Fig. 12.3. At 88·10 K there is a transition from the δ to the α form. The transition is rapid and the δ phase can only be supercooled by a few tenths of a kelvin. At 49·43 K, the α form is in equilibrium with the β. Here, the transition is slow, taking many hours at 40 K, so that the α form may easily be retained on cooling rapidly. The β phase persists to absolute zero; but, on cooling the α, it undergoes another transition at 30·29 K to the γ phase. The entropy of the α

[5] Of course, if a minimum is too narrow or its potential barrier too low, it cannot give rise to a stable (bound) state of the system because of the fundamental confinement limitation indicated by the uncertainty principle.

[6] F. Lange, *Z. Phys. Chem.*, **110**, 343, 1924.

[7] K. K. Kelley, *J. Amer. Chem. Soc.*, **51**, 1400, 1929.

[8] E. D. Eastman and W. C. McGavock, *J. Amer. Chem. Soc.*, **59**, 145, 1937.

[9] C. C. Stephenson and W. F. Giauque, *J. Chem. Phys.*, **5**, 149, 1937.

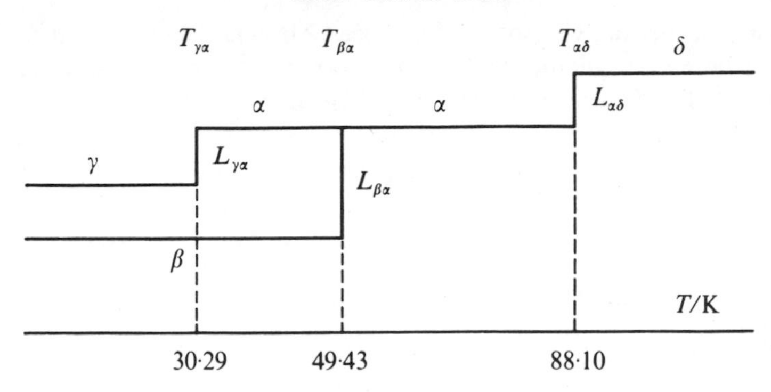

Fig. 12 . 3. The allotropic transitions in phosphine.

phase at $T_{\beta\alpha}$ may be calculated by two paths:

(1)
$$\int_0^{T_{\gamma\alpha}} \frac{C_\gamma}{T}\,\mathrm{d}T + \frac{L_{\gamma\alpha}}{T_{\gamma\alpha}} + \int_{T_{\gamma\alpha}}^{T_{\beta\alpha}} \frac{C_\alpha}{T}\,\mathrm{d}T$$

(2)
$$\int_0^{T_{\beta\alpha}} \frac{C_\beta}{T}\,\mathrm{d}T + \frac{L_{\beta\alpha}}{T_{\beta\alpha}}.$$

The results of Stephenson and Giauque's measurements are summarized in Table 12.1. The agreement is very good.

Table 12.1 Entropy of α-Phosphine at 49·43 K

Route 1			*Route 2*		
Contribution		ΔS/J K^{-1} mol^{-1}	Contribution		ΔS/J K^{-1} mol^{-1}
0–15 K	e	2·070	0–15 K	e	1·415
15–30·29 K	m	9·150	15–49·43 K	m	16·920
$\gamma \rightarrow \alpha$	m	2·710	$\beta \rightarrow \alpha$	m	15·730
30·29–49·43 K	m	20·095			
	Total: 34·03			Total: 34·07	

e = extrapolated; m = measured.

12.5. Glasses

When the temperature of a supercooled liquid is lowered, it eventually either passes over spontaneously into the solid state or becomes glass-like. The transition to the glass is not discontinuous but takes place over a small range of temperature as a result of a rapidly increasing viscosity. The viscosity becomes so high that flow rates become negligible and the substance behaves in many ways like a solid. However, there is no discontinuous change in order and the structure of the glass is much closer to that of a liquid than a true solid. Now, whereas the supercooled liquid is a well defined metastable state, the glass is often not in a state of equilibrium at all as has been demonstrated by measurements on glycerine.

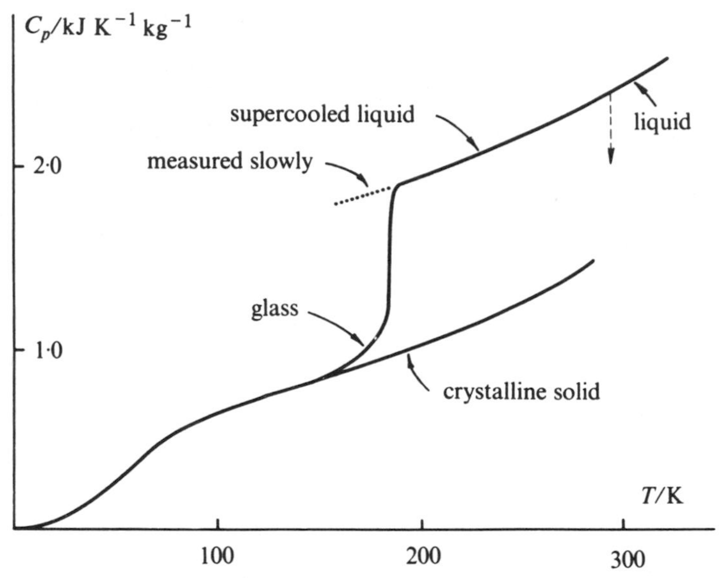

Fig. 12 . 4. The specific heat of glycerol.

The normal freezing point of glycerine is 291 K, but if cooled carefully it remains in the liquid state. At about 180 K the supercooled liquid passes into the glass. The measured heat capacities are shown in Fig. 12.4[10]. Now, the difference in entropy of the solid and liquid at 291 K is given by the latent heat, and by integrating the difference of the heat capacities of the solid and liquid/glass states downwards from that temperature, the difference in the entropies at lower temperatures is found. When this is done (Fig. 12.5), it is found that there is a large entropy difference between the solid and glass at absolute zero in apparent contradiction of the third law.

[10] From measurements of G. E. Gibson and W. F. Giauque, *J. Amer. Chem. Soc.*, **45**, 93, 1923; and of F. E. Simon and F. Lange, *Zeit. für Phys.*, **38**, 227, 1926.

The explanation of this is that the cold glass is not in thermodynamic equilibrium for the following reason. In the liquid state there is no long-range order in the molecular arrangement, but there is short range order, which varies in extent and nature with temperature. As the temperature is changed in the liquid, the adjustment to the equilibrium configuration is able to proceed rapidly; but in the glass, because of the very high viscosity, this is not the case. The time for relaxation of the structure to the equilibrium configuration

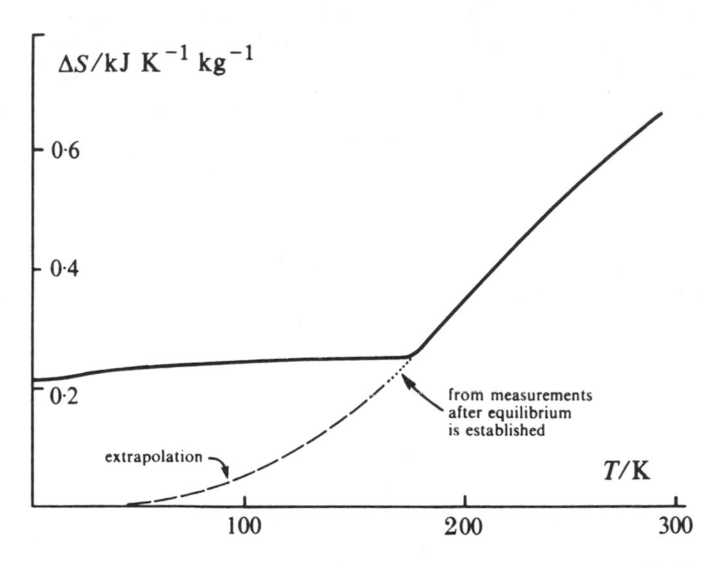

Fig. 12.5. The entropy difference between supercooled and crystalline glycerol.[11]

becomes so long, that especially at lower temperatures, equilibrium may never be reached. In this case, a heat capacity measured at a particular temperature will not correspond to an equilibrium state of the system, and also, in the absence of equilibrium, the third law cannot apply at absolute zero.

The truth of this explanation has been supported by heat capacity measurements on the glass close to the transition where the relaxation times are not excessively long[12]. By waiting for periods up to one week for equilibrium to be established, the true heat capacity

[11] After F. Simon, *Z. Anorg. Allgem. Chem.*, **203**, 219, 1931.

[12] A. G. Oblad and R. F. Newton, *J. Amer. Chem. Soc.*, **59**, 2495, 1937.

curve has been followed down to about 165 K (dotted curve of Fig. 12.4). This corresponds to an entropy difference between the solid and the glass which continues to fall (dotted curve of Fig. 12.5), and suggests that if the equilibrium configuration could be followed to lower temperatures then the entropy difference would indeed become zero at absolute zero (dashed extrapolation of Fig. 12.5).

It is worth pointing out that although a glass is not in an equilibrium state (the potential is not a minimum), certain consequences of the third law still apply. For example, thermal expansion is associated with thermal excitation of the molecules about their frozen-in positions. This aspect of the system remains in equilibrium to absolute zero and the third law therefore applies to show that the expansion coefficient goes to zero.

12.6. The Equilibrium Constant

According to the result of section 11.6, the temperature derivative of the equilibrium constant for a perfect gas reaction is related to the heat of reaction by the equation

$$RT^2 \frac{\mathrm{d}}{\mathrm{d}T} \ln K_p = Q^* = \Delta H_m. \tag{12.6}$$

Now

$$\frac{\mathrm{d}}{\mathrm{d}T} Q^* = \frac{\mathrm{d}}{\mathrm{d}T} \Delta H_m = \Sigma \nu_k C_{mpk}. \tag{12.7}$$

If, therefore, the heat of reaction is determined at one temperature a knowledge of the heat capacities of the reactants is sufficient to enable one to calculate it at other temperatures by integration of (12.7). This, in turn, allows (12.6) to be integrated to give the equilibrium constant at various temperatures *provided it is known at one temperature.* Thus, without the third law and with purely thermal measurements, the temperature variation of K_p may be found, but not its absolute value. The third law was originally put forward by *Nernst* to provide a basis for calculation of the absolute value of the reaction constant using only calorimetric data. His statement was : *If a chemical change takes place between pure crystalline solids at the absolute zero, there is no change of entropy.*

This clearly contains much less than the Simon statement. With the aid of the third law, the equilibrium constant may be calculated as follows.

According to (11.72) the equilibrium constant is related to the molar partial potentials at the standard pressure by the equation

$$RT \ln K_p = -\Sigma \nu_i \, \mu_{0i}. \tag{12.8}$$

Now

$$\begin{aligned} \Sigma \nu_i \, \mu_{0i} &= \Sigma \nu_i (H_{m0i} - TS_{m0i}) \\ &= \Sigma \nu_i \, (H_{mi} - TS_{m0i}) \end{aligned}$$

since the enthalpy of a perfect gas is independent of pressure. Or

$$\Sigma \nu_i \, \mu_{0i} = Q^* - T \, \Sigma \nu_i S_{m0i}. \tag{12.9}$$

Now the first term on the right of (12.9) is the heat of reaction and may be determined directly by a calorimetric experiment. The entropies at the standard pressure appearing in the second term are given by

$$S_0 = S\,(p_0, T) = S\,(p_0, 0) + \int_0^T \mathrm{d}S$$

where the integral may be evaluated on purely thermal data, namely from (*a*) the heat capacities of the gases and their condensed phases down to absolute zero, and (*b*) the latent heats of the transitions between phases. According to the Nernst statement,

$$\Sigma \nu_k \, S_k(p_0, 0) = 0$$

while according to the Simon statement, all the $S_k(p_0, 0)$ are identically zero, so that only the integrals remain in the summation of (12.9). Of course, the heat capacities cannot be measured to absolute zero but it is sufficient to measure them to a temperature low enough to allow a safe extrapolation. Substitution in (12.8) then yields K_p.

The third law has somewhat outlived its original purpose, for the development of statistical mechanics has made it possible to calculate the absolute entropy of a system, without reference to calorimetric measurements, when its microscopic nature is known. It is, however, still very useful in indicating the necessary limiting behaviour of various quantities at low temperatures, and it provides a useful check both for experiments and for theories.

Appendix. Magnetic Energy

There are two forms of the first law which may be found in treatments of magnetic effects, and there has been considerable confusion as to which of them is 'correct'. It is the purpose of this appendix to explain the difference between them.

In section 3.5.4, we showed that the work done on a magnetic material in the process of changing its magnetization is

$$đW = \mathbf{B}.\mathrm{d}\mathbf{m} \tag{A.1}$$

which gives the first law in the form

$$\mathrm{d}U = T\,\mathrm{d}S + \mathbf{B}.\mathrm{d}\mathbf{m}. \tag{A.2}$$

The alternative form for the work done in magnetizing a body is

$$đW' = -\,\mathbf{m}.\mathrm{d}\mathbf{B} \tag{A.3}$$

giving for the first law

$$\mathrm{d}U' = T\,\mathrm{d}S - \mathbf{m}.\mathrm{d}\mathbf{B}. \tag{A.4}$$

Does U or U' represent the true energy of the system?

Let us first show how $đW'$ is derived. We shall again suppose that the magnetic field with which we magnetize the body is produced by a solenoid, but this time, we shall keep the current through the solenoid constant and calculate the mechanical work done as we magnetize the body by moving it into the field.

The field induces a dipole moment in the material, **M**, and when this is in a region of non-uniform field, there is a net force

$$\mathbf{F} = (\mathbf{m}.\boldsymbol{\nabla})\,\mathbf{B}.$$

The work done in a small displacement **ds** is

$$đW' = -\mathbf{F}.\mathrm{d}\mathbf{s} = -\mathbf{m}.\mathrm{d}\mathbf{B}$$

which is (A.3) and gives the first law as (A.4).

Both these sets of equations are correct. The difference is that in the second case some of the mechanical work involved in moving the material into the field is done *on the source of the field.* We may calculate what this amounts to by finding the e.m.f. induced in the solenoid by the moving body. Using the same notation as in section 3.5.4, the mutual inductance of the solenoid and an elementary current loop in the material is

$$L_{12} = \mathbf{b}.\mathbf{a}$$

and the flux threading the solenoid due to i_2 flowing in this loop is

$$d\Phi = i_2\mathbf{a}.\mathbf{b} = \mathbf{m}'.\mathbf{b}$$

That due to the whole body is

$$\Phi = \Sigma(\mathbf{m}'.\mathbf{b}).$$

As the body is moved into the solenoid, both $\mathbf{m}'$ and $\mathbf{b}$ will change. The induced e.m.f. is therefore

$$\mathscr{E} = \dot{\Phi} = \frac{d}{dt}\ \Sigma(\mathbf{m}'.\mathbf{b})$$

and the work done *by the battery* in a small change

$$đW = i_1\ \mathscr{E}\ dt = d(\Sigma(\mathbf{m}'.\mathbf{B}))$$

where $\mathbf{m}$ is the total magnetic moment of the specimen.
which, if the field is uniform over the body, becomes

$$đW = d(\mathbf{m}.\mathbf{B}),$$

Thus, the *net* work done *on the magnetic material* is

$$đW = d(\mathbf{m}.\mathbf{B}) - \mathbf{m}.d\mathbf{B} = \mathbf{B}.d\mathbf{m}$$

which is identical to (A.1).

Thus, there is no inconsistency between the two sets of equations. They refer to different systems. (A.1) and (A.2) apply to the magnetic material only, while (A.3) and (A.4) include the source of the field as part of the system. Both are correct, and, obviously, they will lead to identical physical results.

It should be noted that the effect of including the source of field is to make $\mathbf{B}$ the extensive and $\mathbf{m}$ the intensive variable, and to exchange the internal energy with the enthalpy and the Helmholtz function with the Gibbs function. For example, defining the internal energy through (A.2), the enthalpy becomes

$$H(S, \mathbf{B}) = U - \mathbf{m}.\mathbf{B}$$

and

$$dH = T\,dS - \mathbf{m}.d\mathbf{B} = dU'.$$

while defining it from (A.4), the enthalpy becomes

$$H'(S, \mathbf{m}) = U' = \mathbf{m}.\mathbf{B}$$

and

$$\mathrm{d}H' = T\,\mathrm{d}S + \mathbf{B}.\mathbf{dm} = \mathrm{d}U.$$

Generally, in classical thermodynamics, it is more convenient to exclude the source of field from the system and to use equations (A.1) and (A.2); but equations (A.3) and (A.4) are the natural ones to use when the microscopic nature of the system is being considered as is done in statistical mechanics. The natural way to construct the energy of the system in this case is to write

$$U' = U_0 - \Sigma(\mathbf{m}'.\mathbf{B})$$

where U_0 is the internal energy in other than magnetic forms. If the field is uniform over the body, this becomes

$$U' = U_0 - \mathbf{m}.\mathbf{B}$$

and in a small change in the state of the system the change in U' is

$$\mathrm{d}U' = \mathrm{d}U_0 - \mathbf{B}.\mathbf{dm} - \mathbf{m}.\mathbf{dB}.$$

We now examine the significance of the term $\mathbf{B}.\mathbf{dm}$ in (A.5). For the magnetization to change, the microscopic dipoles must change their orientations. That is, they must undergo transitions between their energy levels. The energy changes involved in this will normally be communicated to the lattice in which the dipoles are situated by the emission or absorption of phonons. Thus, when transitions occur, an amount of energy $\mathbf{B}.\mathbf{dm}$ passes from the magnetic part of the system into the non-magnetic. We may therefore write, in general,

$$\mathrm{d}U_0 = T\,\mathrm{d}S + \mathbf{B}.\mathbf{dm}$$

(assuming other forms of work to be negligible), giving

$$\mathrm{d}U' = T\,\mathrm{d}S - \mathbf{m}.\mathbf{dB}.$$

These are just (A.2) and (A.4), and we see that the latter turns out to be the natural energy function in this case.

Useful Data

Fundamental Constants

speed of light	$c = 3{\cdot}00 \times 10^{8}$ m s^{-1}
electronic charge	$e = 1{\cdot}60 \times 10^{-19}$ C
gravitational constant	$G = 6{\cdot}67 \times 10^{-11}$ N m^{2} kg^{-2}
Planck constant	$h = 6{\cdot}63 \times 10^{-34}$ J s
	$h/2\pi = \hbar = 1{\cdot}05 \times 10^{-34}$ J s
Boltzmann constant	$k = 1{\cdot}38 \times 10^{-23}$ J K^{-1}
electron mass	$m_e = 9{\cdot}11 \times 10^{-31}$ kg
proton mass	$m_p = 1{\cdot}67 \times 10^{-27}$ kg
unified atomic mass constant	$m_u = 1{\cdot}66 \times 10^{-27}$ kg
Avogadro constant	$N_A = 6{\cdot}02 \times 10^{23}$ mol^{-1}
Faraday constant	$F = 9{\cdot}65 \times 10^{4}$ C mol^{-1}
gas constant	$R = 8{\cdot}31$ J K^{-1} mol^{-1}
permeability of vacuum	$\mu_0 = 4\pi \times 10^{-7}$ H m^{-1}
permittivity of vacuum	$\epsilon_0 = 10^{-9}/36\pi$ F m^{-1}
Stefan–Boltzmann constant	$\sigma = 5{\cdot}67 \times 10^{-8}$ W m^{-2} K^{-4}

Other Useful Data

angström	1 Å $= 10^{-10}$ m
electronvolt	1 eV $= 1{\cdot}60 \times 10^{-19}$ J
acceleration of free fall	$g = 9{\cdot}81$ m s^{-2}
molar volume of ideal gas at s.t.p.	$= 2{\cdot}24 \times 10^{-2}$ m^{3}
density of mercury at 15°C	$= 13{\cdot}6 \times 10^{3}$ kg m^{-3}
standard atmosphere	1 atm $= 1{\cdot}01 \times 10^{5}$ Pa
millimetre of mercury	1 mmHg $= 133$ Pa
specific heat of water at 15°C	$= 4{\cdot}19 \times 10^{3}$ J K^{-1} kg^{-1}

Problems

The first part of the problem numbers refers to the chapter with which the problem is most closely associated.

1.1 w is a function of three variables x, y, and z. Prove that

(a)
$$\left(\frac{\partial w}{\partial x}\right)_{y,\,z}=1\bigg/\left(\frac{\partial x}{\partial w}\right)_{y,\,z}$$

(b)
$$\left(\frac{\partial w}{\partial x}\right)_{y,\,z}\left(\frac{\partial x}{\partial z}\right)_{w,\,y}\left(\frac{\partial z}{\partial w}\right)_{x,\,y}=-1$$

which are the reciprocal and reciprocity theorems for functions of three variables.

1.2 A and B are both functions of the variables x and y, and $A/B=C$. Show that

$$\left(\frac{\partial x}{\partial y}\right)_C=\frac{\left(\dfrac{\partial(\ln B)}{\partial y}\right)_x-\left(\dfrac{\partial(\ln A)}{\partial y}\right)_x}{\left(\dfrac{\partial(\ln A)}{\partial x}\right)_y-\left(\dfrac{\partial(\ln B)}{\partial x}\right)_y}.$$

2.1 Can isotherms for different temperatures intersect?

2.2 A constant volume gas thermometer contains a gas whose equation of state is

$$\left(p+\frac{a}{V_m^2}\right)(V_m-b)=RT,$$

and another, of identical construction, contains a different gas which obeys the ideal gas law, $pV_m=RT$. The thermometers are calibrated at the ice and steam points. Show that they will give identical values for a temperature. (Assume that the thermometers are constructed so that all the gas is at the temperature being measured.)

2.3 A constant volume gas thermometer of volume 1×10^{-3} m^3 contains 0·05 mol of a gas. It is assumed that the gas obeys the perfect gas law, $pV_m = RT$; but, in fact, 1 mol of the gas obeys the equation

$$\left(p = \frac{0{\cdot}8}{V_m{}^2}\right)(V_m - 3 \times 10^{-5}) = 8{\cdot}2 \times 10^{-5}T$$

where T K is the thermodynamic temperature, and p and V_m are measured in atmospheres and cubic metres respectively. By how much will temperature measurements be in error at 100°C?

3.1 A system consists of a battery of constant e.m.f. $\mathscr{E}$, in series with a capacitor of capacitance C, which is initially uncharged. Find the work required to pass a charge Q through the system.

3.2 A spring obeys Hooke's law: $\mathscr{F} = y(L - L_0)$, where $\mathscr{F}$ is the tensional force, L the length, L_0 the length at zero tension, and y is a constant. Show that the work done in stretching the spring from L_1 to L_2 is $W = \frac{1}{2}y(L_2 - L_1)(L_2 - 2L_0 + L_1)$.

3.3 The equation of state of an ideal elastic substance is

$$\mathscr{F} = KT\left(\frac{L}{L_0} - \frac{L_0^2}{L^2}\right)$$

where K is a constant and L_0, the length at zero tension, is a function of the temperature T only. Calculate the work required to compress the substance reversibly and isothermally from $L = L_0$ to $L = \frac{1}{2}L_0$.

3.4 One mole of an imperfect gas obeys the equation

$$\left(p + \frac{0{\cdot}8}{V^2}\right)(V - 3 \times 10^{-5}) = 8{\cdot}2 \times 10^{-5}T$$

where p, V, and T are measured in atmospheres, cubic metres, and kelvins respectively. Calculate the work required to compress 0·3 mol of this gas isothermally from a volume of 5×10^{-3} m^3 to 2×10^{-5} m^3 at 300 K.

3.5 Iron ammonium alum is a paramagnetic salt which obeys Curie's law reasonably well to below 1 K. Given that $\chi = 0{\cdot}19/T$, calculate the work done in magnetizing 1×10^{-5} m^3 of the salt in a field of induction 1 T at 4·2 K.

3.6 In a certain calorimetric experiment designed to determine the specific heat of copper, 0·1 kg of the metal at 100°C are added to 2×10^{-4} m^3 of water at 15°C which are contained in a thermally insulated vessel of negligible thermal capacity. After the mixture has reached equilibrium, the temperature is found to be 18·8°C. If the heat capacities of copper and water are essentially constant over the temperatures concerned, what is the specific heat of copper?

Does this experiment measure c_p or c_V? Would there be much difference between these quantities for copper under the conditions of the experiment?

3.7 Below 100 K the specific heat of diamond varies as the cube of the absolute temperature: $c_p = aT^3$. A small diamond, weighing 100 mg, is cooled to 77 K by immersion in liquid nitrogen. It is then dropped into a bath of liquid helium at 4·2 K, which is the boiling point of helium at atmospheric pressure. In cooling the diamond to 4·2 K, some of the helium is boiled off. The gas is collected and found to occupy a volume of $2{\cdot}48 \times 10^{-5}$ m^3 when measured at 0°C and 1 atm pressure. What is the value of a in the formula for the specific heat of diamond?

[The latent heat of vaporization of helium at 4·2 K is 21 kJ kg^{-1}.]

3.8 Water flows through the tube shown below at the rate of 1×10^{-5} m^3 s^{-1}, at which rate the flow is streamlined. What are the differences in the heights of the columns of water in the manometers 1, 2, and 3?

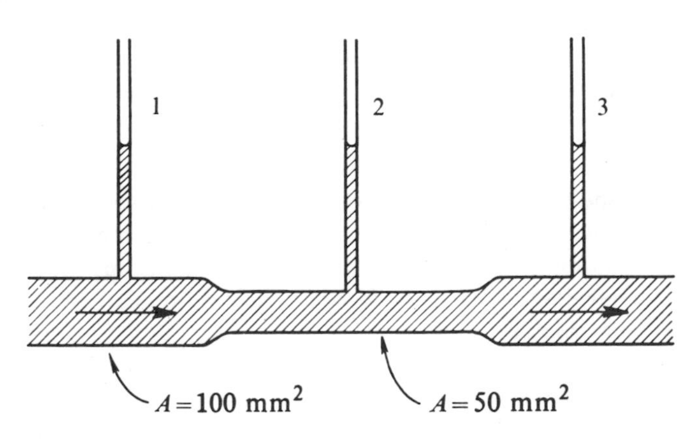

4.1 Prove that if the Kelvin statement of the second law is untrue, then the Clausius statement must also be untrue.

4.2 Prove Carnot's theorem using the Kelvin statement of the second law.

4.3 It has been suggested that a perpetual motion machine could be constructed in the following way. A long vertical tube contains water. At the bottom, the water escapes through a turbine which drives an electrical generator, and the electrical power is used to electrolyse the water to gaseous hydrogen and oxygen. By increasing the height of the column of water, the energy delivered to the turbine by the water may be increased indefinitely so that, by making the column sufficiently tall, more than enough power may be obtained to electrolyse all the water passing through. The excess power may be used to obtain useful work in the surroundings. The whole system is enclosed in another vertical tube and the gases diffuse up this to the top of the water column where they are recombined by a catalyst and replenish the supply of water as quickly as it is used. The device therefore runs indefinitely and provides a perpetual source of power. Where is the fallacy?

4.4 A heat engine operates between a tank containing 1×10^3 m^3 of water and a river at a constant temperature of 10°C. If the temperature of the tank is initially 100°C, what is the maximum amount of work which the heat engine can perform?

4.5 What is the maximum amount of work which may be obtained by operating a heat engine between two beakers of water which are initially at 0°C and 100°C and which both contain 1×10^{-3} m^3 of water?

4.6 An air-conditioning unit consumes 4 kW from the electricity mains and extracts 3 kW of heat from the room it is cooling. Estimate how its efficiency compares to that of an ideal refrigerator.

4.7 Electricity is used to provide domestic hot water at 100°C. The water is supplied at 10°C, and may be heated either with an immersion heater or with an ideal heat pump which extracts heat from the surroundings at 10°C. Compare the power consumptions when (*a*) the immersion heater is used, (*b*) the heat pump supplies heat to the hot water tank so as to keep it at 100°C, (*c*) the heat pump heats 'packets' of water from 10 to 100°C and then adds them to the hot water tank as required.

4.8 A building at a temperature T K is heated by means of a heat pump which uses a river at T_0 K as a source of heat. The heat pump, which may be assumed to have an ideal performance, consumes a constant power W, and the building loses heat to its surroundings at a rate $\alpha(T-T_0)$, where α is a constant. Show that the equilibrium temperature of the building, T_e, is given by

$$T_e = T_0 + \frac{W}{2\alpha}\left\{1 + \left(1 + \frac{4\alpha T_0}{W}\right)^{1/2}\right\}.$$

5.1 Can an isentrope cut an isotherm more than once in a system with two degrees of freedom?

Can you generalize your conclusion so as to be able to say what may happen in a system with more than two degrees of freedom?

5.2 50 kJ of electrical energy are dissipated in 1×10^{-3} m^3 of water. If the water was initially at 15°C, and if that is also the lowest temperature available, what proportion of the energy is reconvertible into work?

5.3 Prove the result of equation (5.15).

5.4 1×10^{-3} m^3 of water is warmed from 20 to 100°C (*a*) by placing it in contact with a large reservoir at 100°C, (*b*) by placing it first in contact with a large reservoir at 50°C until it reaches that temperature, and then in contact with the reservoir at 100°C, and (*c*) by operating a reversible heat engine between it and the reservoir at 100°C. In each case, what are the entropy changes of (i) the water, (ii) the reservoirs, and (iii) the universe?

5.5 Calculate the change in entropy of 1 kg of water when it is heated from 15 to 100°C and completely vaporized.

Does the change in entropy imply any irreversibility in the process?

[The latent heat of vaporization of water at a pressure of 1 atm is $2{\cdot}3 \times 10^6$ J kg^{-1}.]

5.6 Calculate the changes in the entropy of the universe as a result of the following processes:

(*a*) A copper block of mass 400 g and thermal capacity 150 J K^{-1} at 100°C is placed in a lake at 10°C.

(*b*) The same block at 10°C is dropped from a height of 100 m into the lake.

(*c*) Two similar blocks at 100°C and 10°C are joined together.

(*d*) A capacitor of capacitance 1 μF is connected to a battery of e.m.f. 100 V at 0°C.

(*e*) The same capacitor, after being charged to 100 V is discharged through a resistor at 0°C.

(*f*) One mole of a gas at 0°C is expanded reversibly and isothermally to twice its initial volume.

(*g*) One mole of a gas at 0°C is expanded reversibly and adiabatically to twice its initial volume.

5.7 Prove that the change of entropy of the universe when m kg of ice at T_1 K are added to an equal mass of water at T_2 K and the mixture is allowed to reach equilibrium is

$$m\left[c_p^{\mathrm{I}} \ln\frac{273}{T_1}+\frac{L}{273}+c_p^{\mathrm{W}} \ln\left(\frac{T_3^2}{273\,T_2}\right)\right]$$

where c_p^{I} and c_p^{W} are the specific heats of ice and water respectively (assumed constant), L is the latent heat of fusion of ice at 273 K, and T_3 is defined by

$$2\,c_p^{\mathrm{W}}\,T_3=273(c_p^{\mathrm{W}}-c_p^{\mathrm{I}})+T_2\,c_p^{\mathrm{W}}+T_1\,c_p^{\mathrm{I}}-L.$$

5.8 A bath is equipped with two taps, one of which supplies hot water at a temperature of 330 K, and the other cold water at 290 K, which is also the temperature of the surrounding room. The taps are turned on and allowed to deliver 0·1 m^3 of water each. The result, after a certain amount of cooling has occurred, is a bath full of warm water at 305 K. Calculate the amount by which the entropy of the universe increases in the process.
[Ignore the thermal capacity of the bath itself.]

5.9 In a certain experiment, 5 g of liquid helium at 0·5 K are to be cooled further by bringing them into contact with 100 g of a paramagnetic salt which is initially at a lower temperature T_{S}. The specific heat of the liquid may be written $c_{\mathrm{L}}=aT^3$, where $a=20$ J kg^{-1} K^{-4}, and that of the salt may be written $c_{\mathrm{S}}=bT^{-2}$, where $b=0{\cdot}1$ J kg^{-1} K. If the final equilibrium temperature is 0·4 K, calculate the initial temperature of the salt and the net increase in entropy during the process. (Assume the mixture to be mechanically and thermally isolated from the rest of the universe.)

5.10 Three identical bodies of constant thermal capacity C, are initially at temperatures of 300, 300, and 100 K. If no work or heat is available from other sources, then, by operating heat engines between the bodies, (a) what is the maximum amount of work which may be obtained, and (b) what is the highest temperature to which any one of the bodies may be raised?

5.11 The Kapitza helium liquefier uses the following cycle in its first stage. Helium is compressed to 30 atm at 300 K, and passes through a heat exchanger where it is cooled at constant pressure to 10 K. It then expands approximately isothermally in a small expansion engine (doing mechanical work), and is passed back through the heat exchanger at 1 atm, where it warms up to 300 K and returns to the compressor. The helium does not behave like a perfect gas, but its specific heat at constant pressure is approximately independent of pressure. Show that, if the compressor, engine, and heat exchanger are all as efficient as possible, the cycle is reversible. Hence find the ratio of the heat absorbed at 10 K to the total work done on the helium.

7.1 A piece of rubber is subject to work by hydrostatic pressure and by a tensional force.
(*a*) Construct the expression for dU.
(*b*) Generate the potentials which have as proper variables $(S, V, \mathscr{F})$, $(S, p, \mathscr{F})$, and $(T, p, \mathscr{F})$.
(*c*) Derive the Maxwell relations

$$\left(\frac{\partial T}{\partial V}\right)_{L,\,S} = -\left(\frac{\partial p}{\partial S}\right)_{V,\,L}$$

$$\left(\frac{\partial S}{\partial L}\right)_{T,\,p} = -\left(\frac{\partial \mathscr{F}}{\partial T}\right)_{p,\,L}$$

$$\left(\frac{\partial S}{\partial \mathscr{F}}\right)_{p,\,L} = \left(\frac{\partial L}{\partial T}\right)_{S,\,p}.$$

7.2 A gas with the equation of state $p(V-b)=RT$ flows along a thermally insulated tube in which there is a constriction. On passing the constriction, the pressure drops from p_1 to p_2. If the heat capacity of the gas at constant pressure is constant and the speed of flow away from the constriction is small, find the change in temperature.

7.3 A cylinder contains 0·1 kg of water at 15°C. A piston increases the pressure on the water isothermally from 1 atm to 100 atm. Find (*a*) the work done on the water by the piston, (*b*) the heat removed from the water, (*c*) the change in the internal energy of the water.

What would be the change in temperature of the water if the increase in pressure were made adiabatically?

[For water at 15°C, the cubic expansivity is $1{\cdot}5\times10^{-4}$ K^{-1} and the compressibility, $4{\cdot}9\times10^{-12}$ Pa^{-1}.]

7.4 The free energy of a Debye solid may be written in the form

$$F(T, V) = U_0(V) + Tf(\Theta/T),$$

where $U_0(V)$ is the internal energy at absolute zero for the solid with volume V, and Θ is the Debye temperature, a function of the volume only. Obtain an expression for the pressure, and show that the cubic expansivity β_p is related to the isothermal compressibility κ_T by the formula

$$\beta_p = \frac{\kappa_T \gamma C_v}{V} \quad \text{where} \quad \gamma = -\frac{\mathrm{d}(\ln \Theta)}{\mathrm{d}(\ln V)}.$$

8.1 When a thermally isolated atmosphere in a gravitational field is stirred, a vertical temperature gradient is set up. Show that the gradient is given by

$$\frac{\mathrm{d}T}{\mathrm{d}h} = -\frac{(\gamma-1)\,g}{\gamma r}$$

where g is the acceleration of free fall and r is the gas constant for 1 kg of the gas. Estimate the magnitude of this gradient for the atmosphere near the earth's surface.

8.2 A vertical cylinder is closed at the bottom. Gas is enclosed in the cylinder by a close-fitting but frictionless piston above which there is an evacuated space. The piston is displaced slightly from its equilibrium position and released. Show that the period of the resulting oscillations is $2\pi(h/\gamma g)^{1/2}$, where h is the height of the piston above the closed end when it is in equilibrium. The mass of the gas may be assumed to be negligible in comparison with that of the piston.

8.3 A perfect gas with constant specific heats flows adiabatically through a horizontal tube of varying cross-sectional area. If viscous effects are negligible and the flow is streamlined, show that the velocity of flow, $\mathscr{V}$, and the temperature T of the gas vary along the tube according to the relationship

$$\mathscr{V}^2 + \frac{2\gamma}{\gamma-1}\frac{RT}{M_r} = \text{const.}$$

where R is the gas constant, M_r the relative molecular mass and γ the ratio of the specific heats.

8.4 A compressor takes in air at a pressure of 1 atm and delivers compressed air at 10 atm. It uses 250 W of power and its mechanical efficiency is 65 per cent. If the compression is truly adiabatic, at what rate will it deliver compressed air, and at what temperature?

[Take γ for air as 1·4 and assume that the ambient temperature is 300 K.]

8.5 In a certain compressor a perfect gas at room temperature T_0, and atmospheric pressure p_0, is compressed adiabatically, and is then passed through water-cooled tubes until it eventually emerges at pressure p_1 and temperature T_0. Find an expression for the work required for this process, compared with that which would be needed for a reversible isothermal compression leading to the same result and show that the ratio is not less than unity. Examine also the changes of entropy occurring in the two processes.
(Note that if $a > 1$ and $x > 0$, then $a^x > 1 + x \ln a$.)

8.6 If the coldest available reservoir is a lake at 10°C, what is the maximum amount of useful work which may be obtained from 1×10^3 m^3 of a perfect gas which is initially at 100°C and 10 atm pressure and for which $\gamma = 1{\cdot}5$?

8.7 The principal specific heats of a certain perfect gas are $c_p = 1{\cdot}0$ and $c_V = 0{\cdot}7$ kJ K^{-1} kg^{-1}. In a reversible heat engine the gas is (*a*) heated at constant volume until the pressure is 6/5 of its initial value, (*b*) heated at constant pressure until its volume is 5/4 of its initial value, (*c*) cooled at constant volume until the pressure returns to its initial value, (*d*) cooled at constant pressure until the volume returns to its initial value. Find the greatest possible efficiency of this engine and the ratio of the maximum and minimum temperatures of the gas. How does the efficiency of the engine compare with that of a Carnot engine working between the same extremes of temperature?

8.8 A perfect gas, for which $\gamma = 1{\cdot}5$, is used as the working substance in a Carnot engine operating between reservoirs at 300°C and 50°C. The isothermal process at the hotter reservoir consists of an expansion from a pressure of 10 atm and a volume of 1×10^{-3} m^3 to a pressure of 4 atm and a volume of $2{\cdot}5 \times 10^{-3}$ m^3. (*a*) Between what limits of pressure and volume does it operate at the cold reservoir? (*b*) Calculate the heat taken from the source and the heat rejected to the sink in each cycle and show that the efficiency is indeed $(1 - T_2/T_1)$. (*c*) If the area of the piston with which the volume changes are effected is 0·02 m^2 and a force of 100 N is needed to overcome friction between it and the cylinder, calculate the loss in efficiency due to friction.

8.9 In a four-stroke internal combustion engine, the fuel and air mixture is drawn into the cylinder at a temperature T_1 and compressed adiabatically to its burning temperature T_2. It then burns and expands at such a rate that the temperature is steady during the working stroke. At the end of the working stroke, the exhaust valve opens and the burnt gas is swept out. Assuming that throughout the cycle the mixture behaves like a fixed mass of a perfect gas, show that the efficiency cannot be greater than that of the corresponding three-sided reversible cycle, and prove that the latter is given by

$$\eta = 1 - \frac{1 - T_1/T_2}{\ln(T_2/T_1)}.$$

8.10 A gas obeys the equation $p(V-b)=RT$, and has C_V independent of temperature. Show that (*a*) the internal energy is a function of temperature only, (*b*) the ratio $\gamma = C_p/C_V$ is independent of temperature and pressure, (*c*) the equation of an adiabatic change has the form

$$p(V-b)^{\gamma} = \text{constant}.$$

8.11 Show that for a gas obeying van der Waals' equation: $(p+a/V^2)(V-b)=RT$, $T(V-b)^{R/C_V}$ is constant in a reversible adiabatic expansion provided that C_V is independent of temperature.

8.12 Show that the critical constants of the Dieterici and van der Waals gases (see section 8.3) are

Dieterici $\quad p_c = \dfrac{a}{4b^2}e^{-2}, \quad V_c = 2b, \quad T_c = \dfrac{a}{4Rb}$

Van der Waals $\quad p_c = \dfrac{a}{27b^2}, \quad V_c = 3b, \quad T_c = \dfrac{8a}{27Rb}.$

8.13 According to the *law of corresponding states*, the behaviour of all substances should be the same if the pressure, volume, and temperature are measured in *reduced* units, $\pi = p/p_c$, $\phi = V/V_c$ and $\theta = T/T_c$ where p_c, V_c, and T_c are the critical constants. Show that the reduced forms of the Dieterici and van der Waals equations are

Dieterici $\quad \pi(2\phi - 1) = \theta \exp(2 - 2/\theta\phi)$

Van der Waals $\quad (\pi + 3/\phi^2)(3\phi - 1) = 8\theta.$

8.14 The heat capacity of hydrogen at constant volume between 100 and 150 K is adequately described by the equation

$$C_V/R = 1 + 7\times10^{-3}\,(T/\mathrm{K})$$

where R is the gas constant and T the temperature. If some hydrogen is to be cooled from 150 to 100 K by a reversible adiabatic expansion, by what factor must its volume be increased?

8.15 A metal wire of cross-sectional area 0·85 mm² under a tension of 20 N and at a temperature of 20°C is stretched between two rigid supports 1·2 m apart. If the temperature is reduced by 8°C, what is the final tension assuming that the linear expansivity and the Young modulus remain constant at $1{\cdot}5\times10^{-5}$ K^{-1} and 2×10^{11} Pa respectively?

8.16 An elastic filament is such that when stretched by a force $\mathscr{F}$ at a temperature T, the extension x is given by the equation

$$\mu x = \alpha t + \mathscr{F}$$

where $\mu = \mu_0(1+\beta t)$ and $t = T - T_0$, T_0, α, β, and μ_0 being positive constants. When the filament is maintained at a constant length and heated, its thermal capacity is found to be proportional to temperature, $C_x = AT$. Show that

(*a*) A is independent of x.

(*b*) If the entropy is S_0 when $t=0$ and $x=0$, then

$$S = S_0 + \alpha x - \tfrac{1}{2}\mu_0\,\beta x^2 + At.$$

(*c*) If the filament is heated under no tension, the thermal capacity is

$$C_{\mathscr{F}=0} = \left(A + \frac{\mu_0^2\alpha^2}{\mu^3}\right)T.$$

(*d*) For small extensions under adiabatic conditions, the filament cools and the appropriate elastic modulus is

$$\mu + \frac{\mu_0^2\,\alpha^2}{\mu^2 A}.$$

(*e*) When the adiabatic extension is increased so that $x > \alpha/(\beta\mu_0)$, the filament starts to get warmer.

8.17 A wire is under tension. Show that, if the wire suffers an adiabatic fractional increase in length of Δx, then the increase in temperature is given to first order by

$$\Delta T = -\frac{T}{C_L}\left\{\alpha_0 E - \mathscr{F}\,\frac{\mathrm{d}(\ln E)}{\mathrm{d}T}\right\}\Delta x,$$

where C_L is the heat capacity per unit volume of the wire at constant length, $\mathscr{F}$ is the tension per unit cross-sectional area, E is the Young modulus (assumed independent of $\mathscr{F}$) and α_0 is the linear expansivity at zero tension.

8.18 An atomizer produces minute water droplets of diameter 100 nm. A cloud of droplets at 35°C coalesces to form a single drop of water of mass 1 g. Estimate the temperature of the drop if no heat is lost in the process, given that the surface tension of water decreases approximately linearly with temperature and is 75 mN m^{-1} at 5°C and 70 mN m^{-1} at 35°C.

8.19 A sample of iron ammonium alum, a paramagnetic salt, is to be cooled from 4·0 K to 0·5 K by adiabatic demagnetization. What magnetic field is necessary, given that in this temperature range

$$C/R = 1{\cdot}4 \times 10^{-4} T^3 - 7{\cdot}0 \times 10^{-3} T^{-2}$$

and

$$\chi = 0{\cdot}19/T\ ?$$

(Molar volume of iron ammonium alum $= 2{\cdot}82 \times 10^{-4}$ m^3.)

8.20 In a set of experiments on potassium chrome alum, the following data were obtained

T^*	C_B^*/R	$\left[\dfrac{\partial(S/R)}{\partial T^*}\right]_B$
0·064	0·38	12·0
0·054	0·159	7·3
0·044	0·023	2·7
0·034	0·024	6·1

where T^* is the magnetic temperature, $C_B^* = \mathrm{d}Q_B/\mathrm{d}T^*$ is the heat capacity on the magnetic temperature scale measured by adding known amounts of heat to the salt, and $\partial(S/R)/\partial T^*$ is found from measurements at known temperatures in the helium range. Calculate the corresponding absolute temperatures.

8.21 A very small spherical meteor is at a distance from the sun of 50 sun diameters. Estimate its temperature, given that the surface temperature of the sun is 6000 K.

8.22 A solar furnace uses a perfectly reflecting concave mirror of aperture 0·2 m and focal length 2 m. At the position of the image of the sun is a perfectly absorbing sphere 2 mm in diameter made of

a metal with a specific heat of 400 J K^{-1} kg^{-1} and density 8 mg mm^{-3}. Find (*a*) the initial rate of rise in temperature of the sphere, (*b*) the final temperature attainable.
[Angle subtended at the earth by the sun $= 9 \times 10^{-3}$ rad. Temperature of the sun $= 6000$ K.]

8.23 The temperature of a long metal rod of diameter 2 mm is maintained at 1000°C. It is surrounded by two coaxial cylindrical radiation shields of diameter 4 mm and 6 mm, and of negligible thickness. If the entire space is evacuated and all radiating surfaces are black, calculate the temperature of the outer shield when equilibrium has been established assuming that the energy incident from the surroundings is negligible.

8.24 Heat is generated electrically in a long wire at the rate of 10 W m^{-1}. The wire, which is 2 mm in diameter, can only lose heat by radiation to a thin walled coaxial tube, 30 mm in diameter. This, in turn, is covered with a layer of a bad conductor 70 mm thick, the outside of which is maintained at a temperature of 20°C by cooling water. If the thermal conductivity of the bad conductor is $5 \cdot 0 \times 10^{-2}$ W m^{-1} K^{-1}, what will be the temperature of the wire when conditions are steady? (The surfaces of the wire and tube may be regarded as black.)

8.25 A large storage vessel for liquid oxygen may be considered as a perfectly evacuated spherical Dewar vessel of inner radius 1 m, and outer radius 1·2 m. Treating the walls of the vessel as perfectly black, calculate the rate of loss of oxygen due to radiation.

How would the rate of loss be changed if a spherical copper radiation shield were interposed midway between the inner and outer walls?

[Reflectivity of copper $= 0 \cdot 98$ and may be taken as independent of wavelength and temperature. Temperature of the surroundings $= 300$ K. Normal boiling point of liquid oxygen $= 90$ K. Latent heat of oxygen $= 2 \cdot 4 \times 10^5$ J kg^{-1}.]

8.26 A flat, disc-shaped cavity of radius very much larger than its width, is formed from plane surfaces having emissivities α_1 and α_2. Show that if the two surfaces are maintained at temperatures T_1 and T_2 ($T_1 > T_2$), the net flow of radiation across the cavity is equal to that which would be present if the surfaces were black but had an effective radiation constant

$$\sigma_{\text{eff}} = \frac{\alpha_1 \alpha_2 \sigma}{\alpha_1 + \alpha_2 - \alpha_1 \alpha_2}$$

where σ is Stefan-Boltzmann constant

8.27 A paramagnetic ideal gas obeys Curie's law: $\chi_m = a/T$, where χ_m is the susceptibility and a is a constant. A volume V_0 of the gas is placed in a magnetic field of flux density B_0, which is then reduced adiabatically to zero. How must the volume be changed as a function of field if the temperature of the gas is to remain constant?

8.28 Show that the entropy density of equilibrium radiation is $\frac{4}{3}AT^3$ where A is the constant appearing in equation (8.61).

n identical containers of volume V and with negligible thermal capacities are initially filled with radiation characteristic of the temperatures $T_1, T_2, \ldots, T_n$. Show that if no work or heat is available from other sources, then by operating reversible heat engines between the containers

(*a*) the maximum work which may be extracted is

$$W = AV\left\{\sum_{i=1}^{n} T_i^4 - n^{-1/3}\left(\sum_{i=1}^{n} T_i^3\right)^{4/3}\right\}$$

(*b*) the highest temperature T to which the radiation in any one of the containers may be raised is given by

$$\left[\sum_{i=1}^{n} T_i^3 - T^3\right]^4 = (n-1)\left[\sum_{i=1}^{n} T_i^4 - T^4\right]^3.$$

9.1 A perfect gas has constant principal specific heats and is initially at a pressure p_1 and a temperature T_1. Find its final temperature when it undergoes an expansion to a pressure p_2, (*a*) without change of entropy, (*b*) without change of internal energy, (*c*) without change of enthalpy.

How would such expansions be achieved?

9.2 One mole of a gas whose equation of state is $p(V-b) = RT$ undergoes a free expansion from a volume $2b$ to a volume $4b$. Calculate the change in entropy and the change in temperature.

9.3 One mole of hydrogen occupies a volume of 0·1 m^3 and is at 300 K. One mole of argon also occupies a volume of 0·1 m^3 but is at 400 K. While isolated from their surroundings, each undergoes a

free expansion, the hydrogen to five times and the argon to eight times its initial volume. The two masses are then placed in thermal contact with each other and reach thermal equilibrium. What is the total change in entropy?
[For hydrogen, $C_V = 10$ kJ K^{-1} kg^{-1}. Argon has a relative atomic mass of 40 and $C_V = 0{\cdot}31$ kJ K^{-1} kg^{-1}.]

9.4 Two vessels, A and B, have equal volumes and negligible thermal capacities. They are thermally insulated from one another and from the surroundings, but are connected by a narrow capillary fitted with a tap. Initially, A contains a perfect gas at a pressure p_0 and temperature T_0 and B is evacuated. The tap is opened, and gas flows from A to B until the pressures becomes equal. What is the final pressure, and what are the final temperatures of A and B?

9.5 The equation of state for helium gas may be written in the form $pV = RT(1 + B/V)$ where B is a function of temperature only and has the following values for 1 mol:

T/K	10	20	30	40	50	60	70
$B/10^{-6}$ m^3	$-23{\cdot}3$	$-4{\cdot}0$	$+2{\cdot}4$	$+5{\cdot}6$	$+7{\cdot}6$	$+8{\cdot}9$	$+9{\cdot}8$

Determine (*a*) the Boyle temperature, (*b*) the inversion temperature, (*c*) the temperature drop when helium initially at 20 K and 10 atm pressure expands into a very large empty vessel under conditions of thermal isolation.
[For helium, $C_V = \frac{3}{2} R$.]

9.6 Find the maximum inversion temperature of the gas whose equation of state is

$$\left(p + \frac{x}{TV^2}\right)(V - y) = RT,$$

where x and y are constants.

9.7 Calculate the ratio of the maximum inversion temperature to the critical temperature for (*a*) a Dieterici gas, and (*b*) a van der Waals gas. (See section 8.3, and problem 8.12.)

9.8 A particular gas has $C_p = \frac{5}{2} R$ and an equation of state

$$p(V - b) = RT$$

where b is a constant equal to 20 cm^3 mol^{-1}. How much will it cool in a Joule–Kelvin expansion from a pressure of 100 atm to 1 atm?

9.9 Show that when a Dieterici gas (section 8.3) suffers a Joule–Kelvin expansion in which the pressure drop is small, then there is no change of temperature when

$$\frac{b}{V_m}=1-\frac{RTb}{2a}.$$

9.10 Compressed helium gas enters the final stage of a helium liquefier at 14 K. A fraction α is liquefied and the rest is rejected as gas at 14 K and at atmospheric pressure. Use the following data to determine the input pressure which allows α to take its maximum value, and determine what that value is.

Pressure/atm:	0	10	20	30	40
Enthalpy of gas at 14 K/kJ kg^{-1}:	87·0	78·2	72·8	71·6	72·3

Enthalpy of liquid helium at atmospheric pressure = 1·0 kJ kg^{-1}

9.11 In the Simon expansion method for liquefying helium a thermally isolated vessel is filled with gas under pressure at 10 K, and the gas is then allowed to leak slowly away. Owing to the low density of liquid helium, it is possible to choose the initial conditions in such a way that the vessel ends up full of liquid boiling at 4·2 K at a pressure of 1 atm. Assuming that until it liquefies the helium can be treated as a perfect gas with $C_p=\frac{5}{2}R$ and that its latent heat is 84 J mol^{-1}, find the initial pressure required.
[Hint: Consider only that fraction of the helium which is left behind at the end, and show that its entropy in the initial and final states must be the same.]

9.12 In a helium liquefier, just before the final expansion valve, the pressure of the gas is 25 atm and the temperature is 6·5 K. After the expansion, the pressure is 1 atm and the temperature is 4·2 K. Select the relevant information from the following data to deduce what fraction is liquefied.

$$s_{\text{gas}}\,(25\text{ atm, }6{\cdot}5\text{ K})-s_{\text{liquid}}\,(1\text{ atm, }4{\cdot}2\text{ K})=2{\cdot}9\text{ kJ kg}^{-1}$$
$$u_{\text{gas}}\,(25\text{ atm, }6{\cdot}5\text{ K})-u_{\text{liquid}}\,(1\text{ atm, }4{\cdot}2\text{ K})=18\text{ kJ kg}^{-1}$$
$$h_{\text{gas}}\,(25\text{ atm, }6{\cdot}5\text{ K})-h_{\text{liquid}}\,(1\text{ atm, }4{\cdot}2\text{ K})=19\text{ kJ kg}^{-1}$$

At a pressure of 1 atm, the latent heat is 21 kJ kg^{-1} and the specific heat of the gas at constant pressure is 6·7 kJ K^{-1} kg^{-1} throughout the relevant temperature range.

9.13 Apply the first law to the thermoelectric effects to obtain equation (9.23) taking the 'irreversible' heat terms into account.

9.14 A thermocouple is made of two metals whose Thomson coefficients are each proportional to the absolute temperature. Show that if one junction is kept at 0°C and the other at t°C, the thermoelectric voltage is given by $\mathscr{E} = at + bt^2$, where a and b are constants.

If $\mathscr{E}$ is expressed in μV, $a = 2{\cdot}8$, and $b = 0{\cdot}0060$ for a copper–lead thermocouple, while $a = -38{\cdot}1$ and $b = 0{\cdot}045$ for a constantan-lead couple. Calculate the Peltier coefficient for a copper–constantan junction and the difference between the Thomson coefficients of copper and constantan at 100°C. At what temperature does $\mathscr{E}$ for a copper–constantan couple take its maximum value?

10.1 When lead is melted at atmospheric pressure the melting point is 327·0°C, the density decreases from 1·101 to $1{\cdot}065 \times 10^4$ kg m^{-3}, and the latent heat is 24·5 kJ kg^{-1}. What is the melting point at a pressure of 100 atm?

10.2 Between 700 and 739 K, the vapour pressure p of (solid) magnesium at temperature T is given approximately by

$$\ln(p/\text{mmHg}) = -\frac{1{\cdot}7 \times 10^4}{(T/\text{K})} + 19{\cdot}6.$$

Deduce the average heat of sublimation for this temperature interval.

10.3 In the transition from ferromagnetic to paramagnetic nickel at the Curie point, 631 K, the heat capacity at constant pressure changes by 6·7 J K^{-1} mol^{-1} and the volume expansivity by $5{\cdot}5 \times 10^{-6}$ K^{-1}. The molar volume of nickel is $6{\cdot}6 \times 10^{-6}$ m^3. Calculate the effect of pressure on the Curie point assuming the transition to be of second order.

10.4 Examine the following experimental data for water in the light of equations (10.19) and (10.23).

Temperature/°C	0	5	10	15	20
Vapour pressure/mmHg	4·581	6·536	9·198	12·77	17·51

Temperature/°C	25	30	35	40
Vapour pressure/mmHg	23·73	31·79	42·14	55·29

For water vapour, $c_p = 1{\cdot}9$ kJ K^{-1} kg^{-1}.

10.5 If c_1 and c_2 are the specific heats of a liquid and of its saturated vapour respectively, and l is the specific latent heat of vaporization, show that

$$c_2 - c_1 = \frac{\mathrm{d}l}{\mathrm{d}T} - \frac{l}{T}.$$

What is the physical significance of the fact that c_2 for saturated steam is negative?

10.6 Show that when the saturated vapour of an incompressible liquid is expanded adiabatically, some liquid condenses out if

$$C_p + T\frac{\mathrm{d}}{\mathrm{d}T}\left(\frac{L}{T}\right) < 0$$

where C_p is the heat capacity of the liquid (which is assumed constant) and L the latent heat of vaporization.

10.7 Show that, for a system consisting of two phases in equilibrium,

$$\left(\frac{\partial p}{\partial V}\right)_S = -\frac{T}{C_V}\left(\frac{\mathrm{d}p}{\mathrm{d}T}\right)^2$$

where $\mathrm{d}p/\mathrm{d}T$ is the slope of the phase equilibrium curve.

10.8 It is known that a certain substance consists of a mixture of liquid and vapour below a curve on the p–V diagram. Elsewhere it is a single phase. Near the critical point (p_c, V_c, T_c), the curve is given to a good approximation by the relation

$$p_c - p = a(V_c - V)^2.$$

A theory of the substance predicts that the vapour pressure of the liquid near the critical point is given by

$$p = A \exp\left[-\beta(T_c - T)^{2/3}\right].$$

Calculate the latent heat near the critical point according to this theory. Is the result reasonable?

10.9 The transition curve between normal and superconducting tin is given approximately by

$$(B_c/mT) = 30{\cdot}4\left[1 - \left(\frac{T}{3{\cdot}73}\right)^2\right].$$

The density of tin is $7{\cdot}29 \times 10^3$ kg m^{-3}. What is the difference between the specific heats of normal and superconducting tin at 2 K?

10.10 The high temperature behaviour of iron may be idealized as follows:
Below 900°C and above 1400°C, α-iron is the stable modification and between these temperatures γ-iron is stable. The specific heats of the phases may be taken as constant, being 0·775 kJ K^{-1} kg^{-1} for α-iron and 0·690 kJ K^{-1} kg^{-1} for γ-iron.
What is the latent heat at each transition?

10.11 A long tube is closed at one end and, at the other, is connected to a supply of oxygen at a pressure of 100 atm. A short length of the tube is subjected to a field of 1 T. What is the pressure of the gas in the centre of this region?
[Oxygen is a paramagnetic gas with $\chi_m = 2{\cdot}0 \times 10^{-6}$ at s.t.p.]

11.1 Two non-ideal gases may be mixed reversibly with semi-permeable membranes (Fig. 11.1). When the coupled membranes are moved slowly a small distance δx, the whole system being kept at a constant temperature, an amount of heat δQ is rejected to the surroundings. Show that there will generally be a net force on the coupled membranes and calculate how it varies with temperature.

11.2 A mixture of 0·1 mol of helium ($\gamma = \frac{5}{3}$) and 0·2 mol of nitrogen ($\gamma = \frac{7}{5}$) is initially at 300 K and occupies 4×10^{-3} m^3. It is compressed slowly and adiabatically until its volume is reduced by (*a*) 1 per cent, (*b*) by one-quarter. Discuss whether the changes in pressure and temperature of the system can be adequately described in both cases by some average value of γ. Calculate the final temperature and pressure in case (*a*).

11.3 A liquid is in equilibrium with a gas phase consisting of a mixture of its vapour and of an insoluble gas whose partial pressure is p'. Assuming that the components in the gaseous phase behave like ideal phases, show that the vapour pressure of the liquid, p, is changed by the presence of the insoluble gas, the change, if it is not too large, being given by

$$\frac{\Delta p}{p_0} = \frac{v_L M_r}{RT} p'$$

where p_0 is the vapour pressure in the absence of the gas, and v_L and M_r are the specific volume of the liquid and the relative molecular mass of the vapour.

11.4 An inverted U-tube of cross-sectional area 100 mm^2 has 100 mg of sodium chloride at the bottom of one arm and 5×10^{-6} m^3 of water in the other, the remaining space being filled with water vapour. It is kept at a constant temperature of 20°C and the water vapour condenses on the salt side to form a salt solution. What is the difference in the levels of the liquids in the arms when the system reaches equilibrium?
[Relative molecular mass of sodium chloride = 58·4. Vapour pressure of water at 20°C = 17·5 mm Hg.]

11.5 Sea-water contains about 30 kg m^{-3} of sodium chloride. Estimate the minimum work required to obtain 1 m^3 of pure water from the sea.
[Relative molecular mass of common salt = 58·4.]

11.6 Show that for a perfect gas reaction taking place at constant volume and temperature, the numbers of moles of the constituents which are present at equilibrium are connected by a relationship of the form

$$n_1^{\nu_1} n_2^{\nu_2} \ldots n_c^{\nu_c} = K_n(T, V)$$

where $K_n(T, V)$ is a function of temperature and volume only given by

$$\ln K_n(T, V) = -\frac{1}{RT}\sum \nu_i \mu_{0i}$$

where the μ_{0i} are the Gibbs potentials for one mole of the constituents in a volume V at temperature T.

Show also that

$$\frac{\partial}{\partial T}(\ln K_n)_V = \frac{\Delta U}{RT^2}$$

where ΔU is the heat of reaction at constant volume, and that

$$\frac{\partial}{\partial V}(\ln K_n)_T = \frac{\Delta p}{RT}$$

where Δp is the change in pressure when ν_i moles react at constant volume.

12.1 Show that the e.m.f. of a reversible electric cell becomes constant as $T \to 0$.

12.2 On the basis of the following information, which is partly hypothesis and partly somewhat simplified experimental data, calculate the melting pressure of the helium isotope ^{3}He at 0 K.
(*a*) Between 0 and 10 μK, the specific heat of the solid is very high, but between 10 μK and 1 K, it is much less than that of the liquid.
(*b*) Below 1 K the specific heat of the liquid is proportional to T.
(*c*) The expansivity of both phases may be assumed to be zero.
(*d*) At 0·4 K, the melting pressure p_m is 30 atm and $dp_m/dT=0$. At 0·7 K, p_m is 33 atm.

12.3 For a superconductor, the specific heats in the superconducting and normal states may be written

$$c_s = c_{el} + aT^3$$

and

$$c_n = \gamma T + aT^3,$$

where a and γ are constants. Show that

$$\int_0^1 \frac{c_{el}}{\gamma T}\, dt = 1$$

where $t = T/T_c$ is the reduced temperature.

12.4 According to Debye's theory, the heat capacity of a crystalline solid may be expressed in the form

$$C_V = f(T/\Theta)$$

where Θ is independent of temperature but depends on volume according to the law $\Theta \propto V^{-\gamma}$, γ being a constant. Show that the cubic expansivity β and the isothermal compressibility κ satisfy the relation

$$\beta = \gamma\, C_V\, \kappa / V.$$

Index